LECTURES ON
LINEAR ALGEBRA

LECTURES ON
LINEAR ALGEBRA

Donald S Passman
University of Wisconsin-Madison, USA

World Scientific

NEW JERSEY · LONDON · SINGAPORE · BEIJING · SHANGHAI · HONG KONG · TAIPEI · CHENNAI · TOKYO

Published by

World Scientific Publishing Co. Pte. Ltd.

5 Toh Tuck Link, Singapore 596224

USA office: 27 Warren Street, Suite 401-402, Hackensack, NJ 07601

UK office: 57 Shelton Street, Covent Garden, London WC2H 9HE

Library of Congress Control Number: 2022006322

British Library Cataloguing-in-Publication Data

A catalogue record for this book is available from the British Library.

LECTURES ON LINEAR ALGEBRA

ISBN 978-981-125-484-0 (hardcover)
ISBN 978-981-125-499-4 (paperback)
ISBN 978-981-125-485-7 (ebook for institutions)
ISBN 978-981-125-486-4 (ebook for individuals)

For any available supplementary material, please visit
https://www.worldscientific.com/worldscibooks/10.1142/12793#t=suppl

Printed in Singapore

Preface

These are the notes from a one-year, junior-senior level linear algebra course I offered at the University of Wisconsin-Madison some 40 plus years ago. The notes were actually hand written on ditto masters, run off and given to the class at the time of the lectures. Each section corresponds to about one week or three one-hour lectures.

The course was very enjoyable and I even remember three particular A students, but not their names. One always had his hand raised to answer every question I posed to the class. I told him he reminded me of myself when I was an undergraduate. The second had a wonderful sense of humor and always included jokes in each of his homework assignments. The third was a sophomore basketball player. Of course, the U.W. athletics department just couldn't accept the fact that he was a strong math student.

In those years, I believed that elementary row and column operations should not be part of such a course, but when I began to translate these notes into tex, I relented. So I added material on this topic. I also added a brief final chapter on infinite dimensional vector spaces, including the existence of basis and dimension. The proofs here might be considered a bit skimpy.

Most undergraduate linear algebra courses are not as sophisticated as this one was. Most math graduate students are assumed to have had a good course in linear algebra, but sadly many have not. Reading these notes might be an appropriate way to fill in the gap.

D. S. PASSMAN
MADISON, WISCONSIN
SAN DIEGO, CALIFORNIA
FEBRUARY 2021

Contents

CHAPTER I

Vector Spaces

1.   Fields

Linear algebra is basically the study of linear equations. Let us start with a very simple example like $4x + 3 = 5$ or more generally $ax + b = c$. Here of course x is the unknown and a, b and c are known quantities. Now this is really a simple equation and we all know how to solve for x. But before we seek the unknown, perhaps we had better be sure we know the knowns.

First, a, b and c must belong to some system S with an arithmetic. There is an addition denoted by $+$ and a multiplication indicated by juxtaposition, that is the elements to be multiplied are written adjacent to each other. For example, S could be the set of integers, rational numbers $\mathbb{Q}$, real numbers $\mathbb{R}$ or complex numbers $\mathbb{C}$. But, as we will see, the set of integers is really inadequate.

Let us now try to solve for x. Clearly $ax + b = c$ yields $ax = c - b$ and then $x = (c - b)/a$. We see that in order to solve even the simplest of all equations, S must also have a subtraction (the opposite of addition) and a division (the opposite of multiplication). Now the quotient of two integers need not be an integer and we see that the original equation $4x + 3 = 5$ in fact has no integer solution (since of course x must be $1/2$). Thus the integers are not adequate, but it turns out that $\mathbb{Q}$, $\mathbb{R}$ and $\mathbb{C}$ are.

We can now develop three theories of linear algebra, one for each of $\mathbb{Q}$, $\mathbb{R}$ and $\mathbb{C}$. What we would discover is that most theorems proved in one of these situations would carry over to the other two cases and the proofs would be the same. Since it is rather silly to do things three times when one will suffice, we seek the common property of these three sets that make everything work. The common property is that they are *fields*.

Formally, a field F is a set of elements with two operations, addition and multiplication, defined that satisfy a number of axioms. First there are the axioms of addition.

A1. For all $a, b \in F$, we have

$$a + b \in F \qquad\qquad \text{(closure)}$$

A2. For all $a, b, c \in F$, we have

$$a + (b + c) = (a + b) + c \qquad \text{(associative law)}$$

A3. For all $a, b \in F$, we have

$$a + b = b + a \qquad\qquad \text{(commutative law)}$$

A4. There exists a unique element $0 \in F$, called zero, with the property that for all $a \in F$

$$a + 0 = 0 + a = a \qquad \text{(zero)}$$

A5. For each $a \in F$ there is a unique element $-a \in F$, its negative, such that

$$a + (-a) = (-a) + a = 0 \qquad \text{(negative)}$$

Let us consider these in more detail. As we said, the system we work in must have an arithmetic and axiom A1 says that we can add together any two elements of F and get an element of F. Thus the sum of two real numbers is real and the sum of two rational numbers is rational. Associativity is a little harder to understand. We already know how to add two elements of F, but how do we add three? How do we make sense out of $a + b + c$? We might first add a to b, find the answer and then add this to c. That is, we find $(a + b) + c$. On the other hand, we might add b to c, find the answer and then add this to a obtaining $a + (b + c)$. Axiom A2 says that either way we do it, we get the same answer. This therefore allows us to define the sum of three elements unambiguously. Finally, it is not hard to show (see Problem 1.4) that associativity allows us to define the sum of any finite number of elements in an unambiguous manner.

Axiom A3 is a nice assumption. We can add two elements without worrying about which one comes first. Commutativity and associativity have many everyday consequences. For example, let us consider a simple problem in column addition.

$$
\begin{array}{r}
1 \\
2 \\
3 \\
4 \\
+\ 5 \\
\hline
\end{array}
$$

We usually start at the top. We add $1 + 2 = 3$, we add this to 3 getting 6, we then add this to 4 getting 10, and finally we add $10 + 5 = 15$. Of course, what we have really done is computed the expression $(((1 + 2) + 3) + 4) + 5 = 15$. Next, we check our arithmetic by adding upwards and obtain $(((5 + 4) + 3) + 2) + 1 = 15$ since $5 + 4 = 9$, $9 + 3 = 12$, $12 + 2 = 14$ and $14 + 1 = 15$. Thus the answers agree. At first glance we think this is obviously what should happen. But at second glance we see that the equality

$$(((1 + 2) + 3) + 4) + 5 = (((5 + 4) + 3) + 2) + 1$$

is in fact miraculous. It is a consequence of the two axioms of associativity and commutativity.

The final two axioms tell us that F is big enough to allow for a solution of certain linear equations, namely equations of the form $x + a = b$. Thus if $a = b$ then axiom A4 says that $x = 0$ is a solution, and if $b = 0$ then A5 says that $x = -a$ works. In the general case the solution is $x = b + (-a)$. Let us check that it works. We have

$$
\begin{aligned}
x + a &= (b + (-a)) + a \\
&= b + ((-a) + a) &&\text{by the associative law} \\
&= b + 0 &&\text{by the definition of } -a \\
&= b &&\text{by the definition of } 0
\end{aligned}
$$

We might observe that a short hand notation for $b + (-a)$ is $b - a$ and thus F has a subtraction. In many ways subtraction is not as nice an operation as addition. Can we say "subtract a and b"? Certainly not. We must know which of a or b comes first and this means that subtraction is not commutative. Can we subtract a column of figures? Again no, since subtraction is not associative.

The axioms for multiplication are similar to those of addition.

M1. For all $a, b \in F$, we have

$$ab \in F \hspace{4cm} \text{(closure)}$$

M2. For all $a, b, c \in F$, we have

$$a(bc) = (ab)c \hspace{3cm} \text{(associative law)}$$

M3. For all $a, b \in F$, we have

$$ab = ba \hspace{3.5cm} \text{(commutative law)}$$

M4. There exists a unique element $1 \in F$, called one, with the property that for all $a \in F$

$$a1 = 1a = a$$

Furthermore, $1 \neq 0$. \hspace{3cm} (one)

M5. For each $a \in F$ with $a \neq 0$ there is a unique element $a^{-1} \in F$, its inverse, such that

$$aa^{-1} = a^{-1}a = 1 \hspace{3cm} \text{(inverse)}$$

Again associativity and commutativity allow us to multiply any finite number of elements of F unambiguously. Moreover the latter two axioms allow us to solve the equation $ax + b = c$ with $a \neq 0$. The solution is $x = a^{-1}(c + (-b))$ which we check as follows. With this x we have

$$
\begin{aligned}
ax + b &= a(a^{-1}(c + (-b))) + b \\
&= (aa^{-1})(c + (-b)) + b && \text{by the associative law} \\
& && \text{for multiplication} \\
&= 1(c + (-b)) + b && \text{by the definition of } a^{-1} \\
&= (c + (-b)) + b && \text{by the definition of 1} \\
&= c + ((-b) + b) && \text{by the associative law} \\
& && \text{for addition} \\
&= c + 0 && \text{by the definition of } -b \\
&= c && \text{by the definition of 0}
\end{aligned}
$$

So the equation has a solution.

Finally, there is one more axiom. Up till now addition and multiplication have been considered separately. This last axiom intertwines the two.

D. For all $a, b, c \in F$ we have

$$
\begin{aligned}
a(b + c) &= ab + ac, \quad \text{and} \\
(b + c)a &= ba + ca \qquad \text{(distributive law)}
\end{aligned}
$$

The effect of this on the study of linear equations will be apparent in the next section. However, there are a few simple consequences that we state below. Note that part (i) involves 0, an additive element, along with multiplication. Thus, it is clear that some relation between addition and multiplication must be used in the proof. Namely, we have to use the distributive law. We will see how it applies early in the following argument.

LEMMA 1.1. *Let F be a field and let $a, b \in F$.*

i. $a0 = 0a = 0$
ii. $(-a)b = a(-b) = -(ab)$

PROOF. (i) To start with we note that $0 + 0 = 0$ and thus the distributive law implies that

$$
a0 = a(0 + 0) = a0 + a0
$$

For convenience, write $c = a0$. Then $c = c + c$ and by adding $-c$ to both sides of this equation we obtain

$$
\begin{aligned}
0 &= c + (-c)   &&\text{by definition of } -c \\
  &= (c + c) + (-c)   &&\text{since } c = c + c \\
  &= c + (c + (-c))   &&\text{by associativity} \\
  &= c + 0   &&\text{by definition of } -c \\
  &= c   &&\text{by definition of } 0
\end{aligned}
$$

Thus $a0 = c = 0$ and the commutative law yields $0a = a0 = 0$.

(ii) We have

$$
\begin{aligned}
(-a)b + ab &= ((-a) + a)b   &&\text{the distributive law} \\
           &= 0b   &&\text{by definition of } -a \\
           &= 0   &&\text{using } (i) \text{ above}
\end{aligned}
$$

Thus $(-a)b$ is an element of F which when added to ab yields 0. By definition of $-(ab)$, we must have $(-a)b = -(ab)$. Similarly we can show that $a(-b) = -(ab)$. $\qquad\square$

We now consider some examples of fields.

EXAMPLE 1.1. The set of all rational numbers, that is ordinary fractions, is a field which we denote by $\mathbb{Q}$.

EXAMPLE 1.2. The set of all real numbers is a field which we denote by $\mathbb{R}$. As is well known, the real numbers can be put into one-to-one correspondence with the points on a line. We start with two points labeled 0 and 1. Once these are fixed, we can then put in all positive and negative integer points by measuring off multiples of the length of the interval $[0, 1]$. Next, by subdividing all the intervals $[0, n]$ into m equal pieces, we can designate all the rational points. Finally, we somehow "fill in all the holes" with the remaining real numbers.

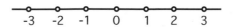

EXAMPLE 1.3. The set of all complex numbers is a field which we denote by $\mathbb{C}$. This field has a property which will be of interest to us. Not only do all linear equations in $\mathbb{C}$ have solutions in $\mathbb{C}$, as we saw above, but in fact all polynomial equations have solutions. Namely let $c_0, c_1, \ldots, c_n \in \mathbb{C}$ with $n \geq 1$ and $c_n \neq 0$, and consider the equation

$$
c_n x^n + c_{n-1} x^{n-1} + \cdots + c_1 x + c_0 = 0
$$

in the unknown x. Then there exists $x \in \mathbb{C}$ that satisfies the equation. This property characterizes what are known as *algebraically closed fields*. At some point, we will see that by assuming that our field F is algebraically closed, we will obtain much nicer theorems on the structure of certain functions called linear transformations.

EXAMPLE 1.4. The above fields all have infinitely many elements. There are however finite fields. The simplest has just two elements. Think about the arithmetic of the ordinary integers and the way even and odd numbers add and multiply. We define a field with two elements "even" and "odd" with addition and multiplication given by

+	even	odd		×	even	odd
even	even	odd		even	even	even
odd	odd	even		odd	even	odd

Here we have described the arithmetic in terms of addition and multiplication tables. It is not hard to check that $F = \{\text{even}, \text{odd}\}$ is indeed a field. Note that even plays the role of 0 and odd plays the role of the element 1.

In the remainder of these lectures the field we work with will play a secondary role. Therefore the reader can, if he or she wishes, assume that F is one of the more familiar examples $\mathbb{Q}$, $\mathbb{R}$ or $\mathbb{C}$. Also the structure and properties of F are assumed to be known. In particular, we will perform the necessary arithmetic operations in F in the familiar fashion, without undo concern with fine points like parentheses or the order of addition. We know that the associative and commutative laws guarantee the validity of these simplifications.

Problems

1.1. Discuss why we abstract the concept of a field.

1.2. Convince yourself that $\mathbb{Q}$, $\mathbb{R}$ and $\mathbb{C}$ are all fields.

Let F be a field and let $a_1, a_2, \ldots, a_n \in F$.

1.3. Consider the column addition of the n field elements listed above. Find the sums adding up and adding down. Show that these sums are the same for $n = 4$. Try the general case.

1.4. Prove that all possible products of $a_1, a_2, \ldots, a_n$ in that order are equal to $(\cdots(((a_1a_2)a_3)a_4)\cdots)a_n$. Try this first for small values of n. This is the *generalized associative law* that allows us to define unambiguously the product $a_1a_2\cdots a_n$.

1.5. Prove that all products of $a_1, a_2, \ldots, a_n$ in any order are equal. This is the *generalized commutative law*.

1.6. Let $a, b, c \in F$ with $a \neq 0$. In the text it was shown that $x = a^{-1}(c + (-b)) = a^{-1}(c - b)$ is a solution of the equation $ax + b = c$. Prove that this is the unique solution.

1.7. Let $a, b, c, d, e, f \in F$. Find the solution to the simultaneous linear equations

$$ax + by = e$$
$$cx + dy = f$$

under the assumption that $ad - bc \neq 0$. What happens if $ad - bc = 0$.

1.8. Let $\mathbb{Q}[i]$ denote the set of all complex numbers of the form $a + bi$ with $a, b \in \mathbb{Q}$ and $i = \sqrt{-1}$. Prove that $\mathbb{Q}[i]$ is a field.

1.9. Let $\mathbb{Q}[\sqrt{2}]$ denote the set of all real numbers of the form $a + b\sqrt{2}$ with $a, b \in \mathbb{Q}$. Prove that $\mathbb{Q}[\sqrt{2}]$ is a field.

1.10. Construct a field containing precisely three elements. You may describe the arithmetic by means of addition and multiplication tables.

2.   Vector Spaces

Let F be a fixed field and let $e, f, g \in F$. Consider the linear equation

$$ex + fy + gz = 0$$

with x, y, z as unknowns. We denote solutions of this, with $x, y, x \in F$, by ordered triples (x, y, z) and let S denote the set of all such solutions.

Suppose (x, y, z) and (x', y', z') are in S. Then by the distributive law and the other axioms of F we have

$$e(x + x') + f(y + y') + g(z + z')$$
$$= (ex + fy + gz) + (ex' + fy' + gz')$$
$$= 0 + 0 = 0$$

Thus $(x + x', y + y', z + z') \in S$. Now this looks like a summation and we are therefore tempted to define an addition in S by

$$(x, y, z) + (x', y', z') = (x + x', y + y', z + z')$$

Now let $(x, y, z) \in S$ and let $a \in F$. Then

$$e(ax) + f(ay) + g(az)$$
$$= a(ex + fy + gz) = a0 = 0$$

and thus $(ax, ay, az) \subset S$. Again this looks like a suitable product and we are therefore tempted to define a scalar multiplication in S, that is a multiplication of an element of S by an element of F, by means of

$$a(x, y, z) = (ax, ay, az)$$

In this way, S is a system that depends on a field F and has defined on it a nice addition and scalar multiplication. Such systems occur frequently in mathematics, as our later examples will indicate, and are known as *vector spaces*.

Formally, a vector space V over F is a set of elements, called *vectors*, with an addition and scalar multiplication satisfying a number of axioms. The axioms of addition say that V has as nice an addition as does the field F. In fact, the axioms are the same. Observe that we use Greek letters to denote vectors.

A1. For all $\alpha, \beta \in V$, we have

$$\alpha + \beta \in V \qquad \text{(closure)}$$

A2. For all $\alpha, \beta, \gamma \in V$, we have

$$\alpha + (\beta + \gamma) = (\alpha + \beta) + \gamma \qquad \text{(associative law)}$$

A3. For all $\alpha, \beta \in V$, we have

$$\alpha + \beta = \beta + \alpha \qquad \text{(commutative law)}$$

A4. There exists a unique vector $0 \in V$, called zero, with the property that for all $\alpha \in V$

$$\alpha + 0 = 0 + \alpha = \alpha \qquad \text{(zero)}$$

A5. For each $\alpha \in V$ there is a unique vector $-\alpha \in V$, its negative, such that

$$\alpha + (-\alpha) = (-\alpha) + \alpha = 0 \qquad \text{(negative)}$$

Note that the same symbol 0 is used to denote the zero in F as well as the zero vector in V. However this almost never causes confusion.

Thus a vector space V satisfies the same addition axioms as does a field. This means, for example, that we can define unambiguously $\alpha_1 + \alpha_2 + \cdots + \alpha_n$ the sum of any finite number of vectors and that this sum is independent of the order in which the summands are written. It means also that we can always solve equations of the form $x + \alpha = \beta$ for $x \in V$. In fact, just about anything true about the addition in fields carries over to V. It is easy to check that the addition defined on S satisfies these axioms.

The axioms of multiplication are a little different. Remember we can only multiply a vector by a field element and the field element always occurs on the left. (This last part is really just a matter of convention.) In particular, there is no commutative law and only one possibility for associativity.

M1. For all $a \in F$ and $\alpha \in V$, we have

$$a\alpha \in V \qquad \text{(closure)}$$

M2. For all $a, b \in F$ and $\alpha \in V$, we have

$$a(b\alpha) = (ab)\alpha \qquad \text{(associative law)}$$

M3. For all $\alpha \in V$, we have

$$1\alpha = \alpha$$

where 1 is the one element of F. $\qquad$ (unital law)

Again associativity guarantees that we can define unambiguously the product vector $a_1 a_2 \cdots a_n \alpha \in V$ with $a_1, a_2, \ldots, a_n \in F$ and $\alpha \in V$. The unital axiom assures us that this multiplication is nontrivial and meaningful. For example if we were to define $a * \alpha = 0$ for all $a \in F$

and $\alpha \in V$, then this multiplication would satisfy all axioms except M3. But certainly this $*$ multiplication is trivial and uninteresting.

Now let us consider the following vector equation

$$ax + \alpha = \beta$$

with $\alpha, \beta \in V$, $a \in F$ and $a \neq 0$. We claim that $x = a^{-1}(\beta + (-\alpha))$ is a solution and we verify this as follows.

$$
\begin{aligned}
ax + \alpha &= a(a^{-1}(\beta + (-\alpha))) + \alpha \\
&= (aa^{-1})(\beta + (-\alpha)) + \alpha && \text{by the associative law} \\
&&& \text{of multiplication} \\
&= 1(\beta + (-\alpha)) + \alpha && \text{by definition of } a^{-1} \\
&= (\beta + (-\alpha)) + \alpha && \text{by the unital law} \\
&= \beta + ((-\alpha) + \alpha) && \text{by the associative law} \\
&&& \text{of addition} \\
&= \beta + 0 && \text{by definition of } -\alpha \\
&= \beta && \text{by definition of } 0
\end{aligned}
$$

Therefore x is indeed a solution. Notice how the unital law gets used above. In the future we denote $\beta + (-\alpha)$ by $\beta - \alpha$.

The remaining two axioms intertwine addition and multiplication.

D1. For all $a \in F$ and $\alpha, \beta \in V$ we have

$$a(\alpha + \beta) = a\alpha + a\beta \qquad \text{(distributive law)}$$

D2. For all $a, b \in F$ and $\alpha \in V$ we have

$$(a + b)\alpha = a\alpha + b\alpha \qquad \text{(distributive law)}$$

Observe that unlike the two distributive laws for fields which are basically the same, the two laws here are really different. This is because scalar multiplication is not only not commutative, but in fact it is not even defined in the other order.

Let us consider some examples of vector spaces.

EXAMPLE 2.1. The simplest nontrivial example of a vector space is F itself. Here vector addition is just ordinary addition in $V = F$ and scalar multiplication is just ordinary multiplication.

EXAMPLE 2.2. Let U be a set and let V be the set of all functions from U into F. If $\alpha, \beta \in V$ and $a \in F$, then we define the functions $\alpha + \beta$ and $a\alpha$ from U to F by

$$(\alpha + \beta)(u) = \alpha(u) + \beta(u)$$
$$(a\alpha)(u) = a(\alpha(u)) \qquad \text{for all } u \in U$$

It is easy to check that with these definitions V is a vector space over the field F.

EXAMPLE 2.3. Now the field $\mathbb{C}$ is endowed with a nice arithmetic and it contains $\mathbb{R}$ as a subfield. Thus there is an addition in $\mathbb{C}$ and a natural multiplication of elements of $\mathbb{C}$ by elements of $\mathbb{R}$. In this way we see easily that $\mathbb{C}$ is a vector space over $\mathbb{R}$.

More generally, if K is any field containing F, then by suitably restricting the multiplication we can make K a vector space over F. This example is not as silly as it first appears. For suppose that we know a good deal about F, but very little about K. Then it is indeed possible to use certain vector space results to deduce information about K. This is in fact a very important tool in the study of fields.

EXAMPLE 2.4. Perhaps the most popular examples of vector spaces are as follows. For each integer $n \geq 1$, let F^n denote the set of n-tuples of elements of F. Thus $F^n = \{(a_1, a_2, \ldots, a_n) \mid a_i \in F\}$ and of course $(a_1, a_2, \ldots, a_n) = (b_1, b_2, \ldots, b_n)$ if and only if the corresponding entries are equal, that is $a_1 = b_1$, $a_2 = b_2$, $\ldots, a_n = b_n$. Now let $\alpha, \beta \in F^n$ with

$$\alpha = (a_1, a_2, \ldots, a_n), \qquad \beta = (b_1, b_2, \ldots, b_n)$$

and let $c \in F$. Then we define

$$\alpha + \beta = (a_1 + b_1, a_2 + b_2, \ldots, a_n + b_n) \in F^n$$

and

$$c\alpha = (ca_1, ca_2, \ldots, ca_n) \in F^n$$

The axioms of F carry over easily to show that F^n is a vector space over F.

EXAMPLE 2.5. Finally, we consider the polynomial ring $F[x]$. This is the set of all polynomials

$$\alpha = \alpha(x) = a_0 + a_1 x + \cdots + a_n x^n$$

with coefficients $a_i \in F$. Observe that α above is also equal to

$$\alpha = a_0 + a_1 x + \cdots + a_n x^n + 0 x^{n+1} + 0 x^{n+2}$$

or in other words, equality of two polynomials means that their nonzero coefficients are the same and that their zero coefficients do not matter. Because of this ambiguity about zero coefficients, it is sometimes simpler to write typical elements $\alpha, \beta \in F[x]$ as

$$\alpha = \sum_{i=0}^{\infty} a_i x^i, \qquad \beta = \sum_{i=0}^{\infty} b_i x^i$$

with the proviso that only finitely many of their coefficients are not zero.

Let $c \in F$. Then with α and β as above, we define

$$\alpha + \beta = \sum_{i=0}^{\infty} (a_i + b_i) x^i$$

$$c\alpha = \sum_{i=0}^{\infty} (ca_i) x^i$$

In this way, $F[x]$ becomes a vector space.

As usual, we can define the degree of a polynomial by

$$\deg 0 = -\infty$$
$$\deg \alpha = \max\{i \mid a_i \neq 0\} \qquad \text{for } \alpha \neq 0$$

Clearly

$$\deg(\alpha + \beta) \leq \max\{\deg \alpha, \deg \beta\}$$

and from this it follows that W_n, the set of all polynomials of degree $\leq n$, is also a vector space under the same operations of addition and scalar multiplication.

Now as we well know, $F[x]$ has a number of additional properties above and beyond its being a vector space. Since we will have need for these later on, it is worthwhile discussing them now. First, polynomials can be multiplied. Roughly speaking to multiply α and β we just distribute the terms, using the fact that $x^i x^j = x^{i+j}$ and then regroup the terms. The answer is $\alpha\beta = \gamma$ where

$$\gamma = \sum_{n=0}^{\infty} c_n x^n$$

and

$$c_n = \sum_{i+j=n} a_i b_j = \sum_{i=0}^{n} a_i b_{n-i} = \sum_{j=0}^{n} a_{n-j} b_j$$

It is easy to check that with this multiplication $F[x]$ satisfies all the axioms of a field with the exception of the existence of multiplicative inverses. In addition, we have

$$\deg \alpha\beta = \deg \alpha + \deg \beta$$

Now by making $F[x]$ somewhat larger, we can in fact get a field. Thus let $F(x)$ denote the set of all rational functions, that is quotients of polynomials with coefficients in F. Then $F(x)$ bears the same relationship to $F[x]$ as $\mathbb{Q}$ does to the set of ordinary integers. Namely, every element of $\mathbb{Q}$ is a quotient of integers and every element of $F(x)$ is a quotient of elements of $F[x]$. Admittedly there are a number of technical details to be checked, but none-the-less with some care we could show that $F(x)$ is indeed a field. Observe that F becomes a subfield of $F(x)$ when we think of each element $a \in F$ as the constant polynomial $a + 0x + 0x^2 + 0x^3 + \cdots$.

Finally, the elements of $F[x]$ can be viewed as functions from F to F. Thus if α is as above and if $c \in F$, then $\alpha(c)$ is given by

$$\alpha(c) = \sum_{i=0}^{\infty} a_i c^i$$

where this is of course a finite sum. It is not hard to see that

$$(\alpha + \beta)(c) = \alpha(c) + \beta(c)$$
$$(\alpha\beta)(c) = \alpha(c)\,\beta(c)$$

We close with an elementary result.

LEMMA 2.1. *Let V be a vector space over the field F. Then for all $a \in F$ and $\alpha \in V$, we have*

    i. $0\alpha = 0$ where the second zero is in V.
    ii. $a0 = 0$ where both zeros are in V.
    iii. $-\alpha = (-1)\alpha$.

PROOF. Now 0 is an additively defined object and the above are multiplicative properties. Thus the proof must be based on the only axioms that intertwine addition and multiplication, namely the distributive laws.

(i) In F we have $0 + 0 = 0$ and thus

$$0\alpha = (0 + 0)\alpha = 0\alpha + 0\alpha$$

Adding $-(0\alpha)$ to both sides yields immediately $0\alpha = 0$.

(ii) In V we have $0 + 0 = 0$ and thus

$$a0 = a(0 + 0) = a0 + a0$$

Adding $-(a0)$ to both sides then yields $a0 = 0$.

(*iii*) We have

$$\alpha + (-1)\alpha = (1)\alpha + (-1)\alpha$$
$$= (1 + (-1))\alpha = 0\alpha = 0$$

Thus adding $-\alpha$ to both sides, we obtain $(-1)\alpha = -\alpha$ and the lemma is proved. $\qquad \square$

Problems

2.1. In the text we showed that a solution of the equation

$$ax + \alpha = \beta$$

with $a \neq 0$ is $x = a^{-1}(\beta - \alpha)$. Prove that this is the unique solution.

2.2. Solve the simultaneous vector equations

$$ax + by = \alpha$$
$$cx + dy = \beta$$

under the assumption that $ad - bc \neq 0$.

2.3. Show that the set of functions from U into F, as described in Example 2.2, is a vector space.

2.4. Prove that F^n is a vector space over the field F. If necessary, try it first for $n = 2$ or 3.

2.5. In F^n define $\alpha_i = (0, 0, \ldots, 1, \ldots, 0)$ where the 1 occurs as the ith entry, and all the other entries are 0. Show that every element of F^n can be written uniquely as a sum

$$\alpha = a_1\alpha_1 + a_2\alpha_2 + \cdots + a_n\alpha_n$$

for suitable $a_i \in F$.

2.6. Clearly $\mathbb{R}^2$ is the ordinary Euclidean plane. Let $\alpha, \beta \in \mathbb{R}^2$. Describe geometrically the quadrilateral with vertices 0, α, $\alpha + \beta$ and β. Describe geometrically the relationship between $a\alpha$ and α for any $a \in \mathbb{R}$.

2.7. Let $\alpha_1 = (1, 1, 2)$, $\alpha_2 = (0, 1, 4)$ and $\alpha_3 = (1, 0, -1)$ be vectors in $\mathbb{Q}^3$. Show that every element $\alpha \in \mathbb{Q}^3$ can be written uniquely as a sum $\alpha = a_1\alpha_1 + a_2\alpha_2 + a_3\alpha_3$ for suitable $a_1, a_2, a_3 \in \mathbb{Q}$.

2.8. Can we find finitely many polynomials $\alpha_1, \alpha_2, \ldots, \alpha_n \in F[x]$ with the property that every element α in $F[x]$ is of the form $\alpha = a_1\alpha_1 + a_2\alpha_2 + \cdots + a_n\alpha_n$ for suitable $a_i \in F$?

2.9. Verify the assertions made in Example 2.5. In particular, check that polynomial multiplication is associative and check that $\deg \alpha\beta = \deg \alpha + \deg \beta$. Also prove that for $a \in F$, $(\alpha + \beta)(a) = \alpha(a) + \beta(a)$ and $(\alpha\beta)(a) = \alpha(a) \cdot \beta(a)$.

2.10. Let α be a nonzero polynomial in $F[x]$ and let $a \in F$ with $\alpha(a) = 0$. Show that $\alpha = (x - a)\beta$ for some $\beta \in F[x]$. Suppose further that $\alpha = (x - a_1)(x - a_2) \cdots (x - a_n)$. Show that $a = a_i$ for some i.

3. Subspaces

Perhaps the subtitle of this section should be "new spaces from old". In mathematics, as soon as an object is defined, one seeks ways of constructing new objects from the original and the obvious first step is to look inside the original for subobjects. Thus we consider subfields of fields, like the reals inside the complex numbers, and of course we consider subsets of sets.

Let V be a vector space over F and let W be a subset of V. Then W is said to be a *subspace* of V if W is a vector space in its own right with the same addition and scalar multiplication as in V. Let us make this last statement more explicit. Suppose $\alpha, \beta \in W$ and $a \in F$. Since W is a vector space, we can compute $\alpha + \beta$ and $a\alpha$ in W. Since α and β also belong to $V \supseteq W$, we can also compute $\alpha + \beta$ and $a\alpha$ in V. Well, the fact that the arithmetic is the same in both W and V means precisely that the two sums $\alpha + \beta$ are the same and again that the two products $a\alpha$ are identical.

At this point, we could start listing examples of vector spaces V and subspaces W, but in each case we would have to verify that W satisfies all ten axioms and this is rather tedious. So what we do first is to decide which of these axioms we really have to check. Suppose W is a subspace of V. Then W has a zero element and negatives, and of course the same is true of V.

LEMMA 3.1. *Let W be a subspace of V. Then the zero element of W is the same as that of V, and negatives in W are the same as in V.*

PROOF. Let 0_W be the zero element of W. Then in W we have $0_W + 0_W = 0_W$. But this is the same addition as in V, so viewing this as an equation in V and adding -0_W (the negative in V) to both sides, we get immediately $0_W = 0$, the zero of V.

Let $\alpha \in W$ and let β be its negative in W. Then $\beta + \alpha = 0_W = 0$. Again, this is the same addition as in V, so adding $-\alpha$ to both sides yields $\beta = -\alpha$. $\square$

We can now obtain our simplified subspace criteria.

THEOREM 3.1. *Let W be a subset of a vector space V. Then W is a subspace of V if and only if*

    i. $0 \in W$,
    ii. $\alpha, \beta \in W$ implies that $\alpha + \beta \in W$, and
    iii. $\alpha \in W$ and $a \in F$ imply that $a\alpha \in W$.

PROOF. Suppose first that W is a subspace of V. Then W is a vector space, so W has a zero element 0_W. By the previous lemma,

$0_W = 0$, so $0 = 0_W \in W$. Finally, if $\alpha, \beta \in W$ and $a \in F$, then by the closure axioms for W, and the fact that the arithmetic is the same in W and in V, we have $\alpha + \beta \in W$ and $a\alpha \in W$. Thus (i), (ii) and (iii) are satisfied.

Now suppose W is a subset of vector space V satisfying (i), (ii) and (iii). We show that W is a vector space in its own right using the same arithmetic operations as in V. We have to check the ten axioms and we consider them in four packages.

1. The two closure axioms are of course immediate consequences of (ii) and (iii).

2. By (i), $0 \in W$ so there exists an element of W, namely 0, with the property that $\alpha + 0 = 0 + \alpha = \alpha$ for all $\alpha \in W$. Suppose some second element $\gamma \in W$ satisfies $\alpha + \gamma = \gamma + \alpha = \alpha$ for all $\alpha \in W$. Then since $0 \in W$, we can set $\alpha = 0$ here and get immediately $\gamma = 0 + \gamma = 0$. Thus W has a unique zero element, the zero of V, and the zero axiom is satisfied.

3. Let $\alpha \in W$. Then by our assumption (iii) and Lemma 2.1(iii), $-\alpha = (-1)\alpha \in W$. Thus there exists an element of W, namely $-\alpha$, with the property that $\alpha + (-\alpha) = (-\alpha) + \alpha = 0$. Now suppose there existed some second element $\gamma \in W$ with $\alpha + \gamma = \gamma + \alpha = 0$. Then viewing this as an equation in V and using the uniqueness of negatives in V, we see immediately that $\gamma = -\alpha$. Thus W has unique negatives for all its elements and the negative axiom is satisfied.

4. Finally, there remains the commutative law, the two associative laws, the two distributive laws and the unital law to consider. But we claim that these are in fact obviously satisfied. For example, consider the distributive law $a(\alpha+\beta) = a\alpha+a\beta$ with $\alpha, \beta \in W$. Now if this were an equation with addition and scalar multiplication in V, then there would be no question of its validity. But this is in fact such an equation since the arithmetic in W is precisely the same as in V. Therefore this axiom is satisfied in W and similarly the other ones hold.

Thus W satisfies all ten axioms and we have proved that W is a subspace of V.   □

We now consider our examples.

EXAMPLE 3.1. Let V be a vector space. Then V always has two subspaces namely V itself and $\{0\}$. We usually denote this latter subspace by 0 and thus this overworked symbol now signifies the zero of F, the zero of V, and the zero subspace. Moreover it will also be used to designate certain additional objects later on. Again this usually causes no confusion.

EXAMPLE 3.2. If $V = F[x]$ is the polynomial ring over F in the variable x, then for each integer n, the set of polynomials of degree $\leq n$ is a subspace of V.

EXAMPLE 3.3. Let U be a set and let V be the set of all functions from U to F. As we have seen earlier, V is a vector space over F with addition and scalar multiplication given by

$$(\alpha + \beta)(u) = \alpha(u) + \beta(u)$$
$$(a\alpha)(u) = a{\cdot}\alpha(u) \qquad \text{for all } u \in U$$

Let u_0 be a fixed element of U. Then it is easy to see that the set $W = \{\alpha \in V \mid \alpha(u_0) = 0\}$ is a subspace of V.

EXAMPLE 3.4. Let V be the set of all real valued functions defined on the interval $[0, 1]$. That is, V is the set of all functions from $U = [0, 1]$ to $F = \mathbb{R}$ and hence, as above, V is a vector space over $\mathbb{R}$. Suppose C denotes the set of all continuous functions in V. Then 0 is continuous, and the sum of two continuous functions is continuous. Furthermore, C is closed under scalar multiplication, so C is in fact a subspace of V.

EXAMPLE 3.5. Now let us consider $V = F^n$ for any field F. If W is the set of all such n-tuples whose first entry is zero, then W is clearly a subspace of V.

More generally, suppose we are given a set of *homogeneous linear equations* over F in n unknowns. For the sake of simplicity, we will write down just three such equations.

$$e_1x_1 + e_2x_2 + \cdots + e_nx_n = 0$$
$$f_1x_1 + f_2x_2 + \cdots + f_nx_n = 0$$
$$g_1x_1 + g_2x_2 + \cdots + g_nx_n = 0$$

Observe that the word homogeneous means that zeros occur on the right hand side of the equal sign. Let S denote the set of all n-tuples $(x_1, x_2, \ldots, x_n) \in F^n$ that are solutions to all three equations. We show that S is a subspace of V.

First $0 = (0, 0, \ldots, 0)$ is clearly in S. Now let $\alpha = (x_1, x_2, \ldots, x_n) \in S$, $\beta = (y_1, y_2, \ldots, y_n) \in S$ and let $a \in F$. Then

$$e_1(x_1+y_1) + e_2(x_2 + y_2) + \cdots + e_n(x_n + y_n)$$
$$= (e_1x_1 + e_2x_2 + \cdots + e_nx_n) + (e_1y_1 + e_2y_2 + \cdots + e_ny_n)$$
$$= 0 + 0 = 0$$

Thus $\alpha + \beta$ satisfies the first equation and, in a similar manner, also the other two. In other words, $\alpha + \beta \in S$. In the same way

$$e_1(ax_1) + e_2(ax_2) + \cdots + e_n(ax_n)$$
$$= a(e_1x_1 + e_2x_2 + \cdots + e_nx_n) = a0 = 0$$

so $a\alpha$ satisfies the three equations and $a\alpha \in S$. This implies that S is a subspace of V. Thus we see that the study of subspaces is a necessary ingredient in the study of linear equations.

EXAMPLE 3.6. Suppose $F \subseteq K \subseteq L$ is a chain of three fields, as for example $\mathbb{Q} \subseteq \mathbb{R} \subseteq \mathbb{C}$. Then L is a vector space over F and K is certainly a subspace. Observe that L is also a vector space over field K and this fact turns out to have many interesting consequences.

Now suppose V is a vector space and that we already know a number of subspaces. In the following few examples, we discuss ways of constructing new subspaces from these.

EXAMPLE 3.7. Let $\{W_i\}$ be a family of subspaces of V. Then the intersection $W = \bigcap_i W_i$ is also a subspace, a fact which we now prove. First, $0 \in W_i$ for all i, so $0 \in W$. Second let $\alpha, \beta \in W$ and $a \in F$. Then for each i, we have $\alpha, \beta \in W_i$ so, since W_i is a subspace, we have $\alpha + \beta \in W_i$ and $a\alpha \in W_i$. That is, $\alpha + \beta, a\alpha \in W_i$ for all i, so $\alpha + \beta, a\alpha \in \bigcap_i W_i = W$. Thus W is a subspace of V and it is clearly the largest subspace contained in all W_i.

EXAMPLE 3.8. Let W_1 and W_2 be subspaces of V. Then, as we observed above, $W_1 \cap W_2$ is the largest subspace of V contained in both W_1 and W_2. Now let us turn this problem around and find the smallest subspace which contains both W_1 and W_2. Suppose first that W is any subspace satisfying $W \supseteq W_1, W_2$. If $\alpha_1 \in W_1$ and $\alpha_2 \in W_2$, then $\alpha_1, \alpha_2 \in W$, so $\alpha_1 + \alpha_2 \in W$. Thus W contains the set of all such sums or in other words $W \supseteq W_1 + W_2$ where we define

$$W_1 + W_2 = \{\alpha_1 + \alpha_2 \mid \alpha_1 \in W_1, \alpha_2 \in W_2\}$$

We show now that $W_1 + W_2$ is in fact a subspace of V and this will therefore imply that $W_1 + W_2$ is the smallest subspace containing both W_1 and W_2.

Let $\alpha, \beta \in W_1 + W_2$ and let $a \in F$. Then by definition of $W_1 + W_2$ there exist $\alpha_1, \beta_1 \in W_1$ and $\alpha_2, \beta_2 \in W_2$ with

$$\alpha = \alpha_1 + \alpha_2, \qquad \beta = \beta_1 + \beta_2$$

Then
$$\alpha + \beta = (\alpha_1 + \alpha_2) + (\beta_1 + \beta_2)$$
$$= (\alpha_1 + \beta_1) + (\alpha_2 + \beta_2) \in W_1 + W_2$$
since $\alpha_1 + \beta_1 \in W_1$ and $\alpha_2 + \beta_2 \in W_2$. Also
$$a\alpha = a(\alpha_1 + \alpha_2)$$
$$= (a\alpha_1) + (a\alpha_2) \in W_1 + W_2$$
since $a\alpha_1 \in W_1$ and $a\alpha_2 \in W_2$. Finally, $0 = 0 + 0 \in W_1 + W_2$, so $W_1 + W_2$ is indeed a subspace.

Now every element of $W_1 + W_2$ can be written as a sum of an element of W_1 and one of W_2, but the summands need not be uniquely determined. If it happens that all such summands are unique, then we say that the sum is *direct* and write $W_1 \oplus W_2$ for $W_1 + W_2$. The following lemma yields a simple test for deciding when a sum is direct.

LEMMA 3.2. *Let W_1 and W_2 be subspaces of V. Then $W_1 + W_2$ is a direct sum if and only if $W_1 \cap W_2 = 0$.*

PROOF. Suppose first that $W_1 + W_2$ is a direct sum and let $\alpha \in W_1 \cap W_2$. Then
$$\alpha = \alpha + 0 \in W_1 + W_2 \qquad \text{(viewing } \alpha \in W_1 \text{ and } 0 \in W_2\text{)}$$
and
$$\alpha = 0 + \alpha \in W_1 + W_2 \qquad \text{(viewing } 0 \in W_1 \text{ and } \alpha \in W_2\text{)}$$
Thus, by uniqueness of summands, we must have $\alpha = 0$ and hence $W_1 \cap W_2 = 0$.

Conversely, suppose $W_1 \cap W_2 = 0$ and let
$$\gamma = \alpha_1 + \alpha_2 = \beta_1 + \beta_2 \in W_1 + W_2$$
with $\alpha_1, \beta_1 \in W_1$ and $\alpha_2, \beta_2 \in W_2$. Then
$$\alpha_1 - \beta_1 = \beta_2 - \alpha_2$$
Now the left hand term here is clearly in W_1 and the right hand term is in W_2. Thus this common element is in $W_1 \cap W_2 = 0$. This yields $\alpha_1 - \beta_1 = \beta_2 - \alpha_2 = 0$, so $\alpha_1 = \beta_1$, $\alpha_2 = \beta_2$, and the sum is direct. $\square$

EXAMPLE 3.9. Of course what we can do with two summands, we can do with finitely many. Let $W_1, W_2, \ldots, W_n$ be subspaces of V and set
$$W_1 + W_2 + \cdots + W_n = \{\alpha_1 + \alpha_2 + \cdots + \alpha_n \mid \alpha_i \in W_i\}$$
It follows as above that $W = W_1 + W_2 + \cdots + W_n$ is a subspace of V and in fact it is the smallest subspace containing $W_1, W_2, \ldots$ and W_n.

If, in addition, every element α of W can be written uniquely as $\alpha = \alpha_1 + \alpha_2 + \cdots + \alpha_n$ with $\alpha_i \in W_i$, then again we say that the sum is direct and we write $W = W_1 \oplus W_2 \oplus \cdots \oplus W_n$. It is not hard to show that $W = W_1 + W_2 + \cdots + W_n$ is a direct sum if and only if

$$(W_1) \cap W_2 = 0$$
$$(W_1 + W_2) \cap W_3 = 0$$
$$(W_1 + W_2 + W_3) \cap W_4 = 0$$
$$\cdots\cdots$$
$$(W_1 + W_2 + \cdots + W_{n-1}) \cap W_n = 0$$

Since addition is commutative, this chain of conditions is therefore equivalent to any similar chain of conditions with the subscripts permuted.

EXAMPLE 3.10. Let us change the problem a little. Suppose $\alpha \in V$. What is the smallest subspace containing α? Well obviously any subspace containing α must necessarily contain all scalar multiples of α and thus we define

$$\langle \alpha \rangle = \{a\alpha \mid a \in F\}.$$

Now it is not hard to see that $\langle \alpha \rangle$ is a subspace. In fact, all the proof amounts to is observing that

$$0 = 0\alpha \in \langle \alpha \rangle$$
$$a\alpha + b\alpha = (a + b)\alpha \in \langle \alpha \rangle$$
$$a(b\alpha) = (ab)\alpha \in \langle \alpha \rangle$$

for all $a, b \in F$. Thus $\langle \alpha \rangle$ is a subspace of V and then it is the smallest subspace containing α.

Finally, let $\alpha_1, \alpha_2, \ldots, \alpha_n \in V$. Then clearly the smallest subspace containing all these vectors is $\langle \alpha_1 \rangle + \langle \alpha_2 \rangle + \cdots + \langle \alpha_n \rangle$ which we denote by $\langle \alpha_1, \alpha_2, \ldots, \alpha_n \rangle$. Thus clearly

$$\langle \alpha_1, \alpha_2, \ldots, \alpha_n \rangle = \{a_1\alpha_1 + a_2\alpha_2 + \cdots + a_n\alpha_n \mid a_i \in F\}$$

Problems

3.1. Let $V = F^2$ and let $\alpha = (1, 0)$, $\beta = (0, 1)$ and $\gamma = (1, 1)$ be three vectors in V. By computing both sides of each inequality, show that

$$\langle \alpha \rangle + (\langle \beta \rangle \cap \langle \gamma \rangle) \neq (\langle \alpha \rangle + \langle \beta \rangle) \cap (\langle \alpha \rangle + \langle \gamma \rangle)$$

and

$$\langle \alpha \rangle \cap (\langle \beta \rangle + \langle \gamma \rangle) \neq (\langle \alpha \rangle \cap \langle \beta \rangle) + (\langle \alpha \rangle \cap \langle \gamma \rangle)$$

Thus appropriate analogs of the distributive law do not hold for these operations on subspaces.

3.2. Let $\alpha_1, \alpha_2, \alpha_3, \alpha_4$ be four vectors in V that satisfy

$$\alpha_1 - 3\alpha_2 + 2\alpha_3 - 5\alpha_4 = 0$$

Show that $\langle \alpha_1, \alpha_2, \alpha_3, \alpha_4 \rangle = \langle \alpha_2, \alpha_3, \alpha_4 \rangle$.

3.3. Let W_1 and W_2 be subspaces of V. Show that $W_1 \cup W_2$ is a subspace if and only if $W_1 \subseteq W_2$ or $W_2 \subseteq W_1$. (Hint. If neither of these two inclusions hold, then we can choose $\alpha_1 \in W_1 \setminus W_2$ and $\alpha_2 \in W_2 \setminus W_1$. Consider the element $\alpha_1 + \alpha_2$.)

3.4. Let W_1, W_2 and W_3 be subspaces of V with $W_1 \subseteq W_3$. Prove that

$$(W_1 + W_2) \cap W_3 = W_1 + (W_2 \cap W_3)$$

This is called the *modular law*.

3.5. Let $\alpha = (1, 1, 2), \beta = (0, 1, 3), \gamma = (2, 0, -1) \in \mathbb{Q}^3$. Show that

$$\mathbb{Q}^3 = \langle \alpha \rangle \oplus \langle \beta \rangle \oplus \langle \gamma \rangle$$

3.6. Prove that the criteria given in Example 3.9 for the sum of subspaces $W_1 + W_2 + \cdots + W_n$ to be direct is indeed correct.

In each of the remaining problems decide whether there exist finitely many vectors $\alpha_1, \alpha_2, \ldots, \alpha_m \in V$ with $V = \langle \alpha_1, \alpha_2, \ldots, \alpha_m \rangle$.

3.7. $V = F^n$.

3.8. $V = F[x]$.

3.9. $V = \mathbb{C}$, $F = \mathbb{R}$.

3.10. $V = \mathbb{R}$, $F = \mathbb{Q}$. This requires outside knowledge about the size of the real numbers.

4.   Spanning and Linear Independence

Let us again return to the problem of a single linear equation, say

$$x_1 - x_2 + x_3 = 0$$

over the field $\mathbb{R}$ of real numbers. Now we seek to solve this equation, that is to find all its solutions. As before, let

$$S = \{(x_1, x_2, x_3) \in \mathbb{R}^3 \mid x_1 - x_2 + x_3 = 0\}$$

Then S is a subspace of $\mathbb{R}^3$ and what we want to do is to list all the elements of S. Now $(2, 1, -1) \in S$ and hence so is $(2r, r, -r)$ for every real number r. Thus S has infinitely many elements, so we cannot hope to list them all. Instead, we devise another scheme.

Since $x_3 = x_2 - x_1$ from the above, a typical element of S looks like

$$(x_1, x_2, x_3) = (x_1, x_2, x_2 - x_1) = x_1(1, 0, -1) + x_2(0, 1, 1)$$

Observe that $\beta_1 = (1, 0, -1)$ and $\beta_2 = (0, 1, 1)$ are elements of S and by the above every element of S can be written as a linear sum of β_1 and β_2 with coefficients in $\mathbb{R}$. Moreover, it is easy to see that the coefficients are unique. What we have found is a basis for S.

Let V be a vector space over F. A *basis* $\mathcal{B}$ is a subset $\mathcal{B} = \{\beta_1, \beta_2, \ldots\}$, possibly infinite, of V such that every element of $\alpha \in V$ can be written uniquely as a finite F-linear sum of elements of $\mathcal{B}$. That is, for each $\alpha \in V$, there exist finitely many elements of $\mathcal{B}$, say $\beta_{i_1}, \beta_{i_2}, \ldots, \beta_{i_m}$ such that

$$\alpha = a_1\beta_{i_1} + a_2\beta_{i_2} + \cdots + a_m\beta_{i_m}$$

with $a_i \in F$. Moreover, with the exception of terms with zero coefficients, the choice of $\beta_{i_1}, \beta_{i_2}, \ldots, \beta_{i_m}$ and the a_i are uniquely determined by α. If $\mathcal{B} = \{\beta_1, \beta_2, \ldots, \beta_n\}$ is a finite set, then the above of course simplifies to the statement that every element $\alpha \in V$ can be written uniquely as

$$\alpha = a_1\beta_1 + a_2\beta_2 + \cdots + a_n\beta_n$$

with $a_i \in F$.

Let us consider some examples.

EXAMPLE 4.1. The space S given above has

$$\mathcal{B} = \{\beta_1 = (1, 0, -1), \beta_2 = (0, 1, 1)\}$$

as a basis.

EXAMPLE 4.2. $V = F[x]$ has $\mathcal{B} = \{1, x, x^2, \ldots, x^n, \ldots\}$ as a basis. In other words, every element $\alpha \in F[x]$ can be written as a finite sum $\alpha = \sum_{i=0}^{m} a_i x^i$ and the only ambiguity that occurs is in the number of zero-coefficient terms that we allow as summands.

EXAMPLE 4.3. F^n has $\mathcal{B} = \{\beta_1, \beta_2, \ldots, \beta_n\}$ as a basis where $\beta_i = (0, 0, \ldots, 1, \ldots, 0)$ has a 1 in its ith entry and zeros elsewhere.

EXAMPLE 4.4. If we view $\mathbb{C}$ as a vector space over $\mathbb{R}$, then $\mathbb{C}$ has $\{1, \sqrt{-1}\}$ as a basis.

EXAMPLE 4.5. If we view $\mathbb{Q}[\sqrt{2}]$ as a vector space over $\mathbb{Q}$, then $\mathbb{Q}[\sqrt{2}]$ has $\{1, \sqrt{2}\}$ as a basis.

We should mention here that vector spaces have very many different bases and that there is usually no such thing as a canonical or best basis. In fact, one of the main problems in linear algebra is to find bases in which the behavior of certain objects can be more easily understood. However, generally for two different such objects, we get two different bases.

Now there are really two aspects of the definition of a basis $\mathcal{B}$. First, each element $\alpha \in V$ can be written as a finite F-linear sum of elements of $\mathcal{B}$, the existence part. Second, the coefficients are unique, the uniqueness part. Each of these aspects is interesting in its own right, and we consider them separately.

Let $\mathcal{C} = \{\gamma_1, \gamma_2, \ldots\}$ be a subset of V. We say that $\mathcal{C}$ is a *spanning* set for V, or that $\mathcal{C}$ spans V, if every element $\alpha \in V$ can be written as a finite F-linear sum

$$\alpha = a_1 \gamma_{i_1} + a_2 \gamma_{i_2} + \cdots + a_m \gamma_{i_m}$$

of elements of $\mathcal{C}$. Observe that if $\mathcal{C} = \{\gamma_1, \gamma_2, \ldots, \gamma_n\}$ is a finite set, then $\mathcal{C}$ spans V if and only if $V = \langle \gamma_1, \gamma_2, \ldots, \gamma_n \rangle$.

Again, let $\mathcal{C}$ be as above. We say that $\mathcal{C}$ is *linearly independent* if for every finite subset $\{\gamma_{i_1}, \gamma_{i_2}, \ldots, \gamma_{i_m}\}$ of $\mathcal{C}$ every equation of the form

$$a_1 \gamma_{i_1} + a_2 \gamma_{i_2} + \cdots + a_m \gamma_{i_m} = 0$$

has only the trivial solution $a_1 = a_2 = \cdots = a_m = 0$. In other words, $\mathcal{C}$ is linearly independent if no nontrivial finite F-linear sum of elements of $\mathcal{C}$ can add to zero. If $\mathcal{C}$ is not linearly independent, then we say that it is *linearly dependent*. Note that if $\mathcal{C} = \{\gamma_1, \gamma_2, \ldots, \gamma_n\}$ is a finite set, then $\mathcal{C}$ is linearly independent if and only if the equation

$$a_1 \gamma_1 + a_2 \gamma_2 + \cdots + a_n \gamma_n = 0$$

can hold only for $a_1 = a_2 = \cdots = a_n = 0$. We consider now the relationship between this concept and the uniqueness aspect of bases.

LEMMA 4.1. *Let* $\mathcal{C} = \{\gamma_1, \gamma_2, \ldots\}$ *be a subset of* V. *Then* $\mathcal{C}$ *is linearly independent if and only if every element* $\alpha \in V$ *that can be written as an F-linearly combination of finitely many elements of* $\mathcal{C}$ *can be written so uniquely.*

PROOF. Suppose first that uniqueness holds and let $\langle \gamma_{i_1}, \gamma_{i_2}, \ldots \gamma_{i_m} \rangle$ be a finite subset of $\mathcal{C}$. Say $a_1, a_2, \ldots, a_m \in F$ with

$$a_1 \gamma_{i_1} + a_2 \gamma_{i_2} + \cdots + a_m \gamma_{i_m} = 0$$

Since we know that

$$0\gamma_{i_1} + 0\gamma_{i_2} + \cdots + 0\gamma_{i_m} = 0$$

uniqueness for the vector $\alpha = 0$ implies that $a_1 = a_2 = \cdots = a_m = 0$ and $\mathcal{C}$ is linearly independent.

Conversely suppose that $\mathcal{C}$ is linearly independent and suppose $\alpha \in V$ can be written as

$$b_1 \gamma_{i_1} + b_2 \gamma_{i_2} + \cdots + b_r \gamma_{i_r} = \alpha$$

and

$$c_1 \gamma_{j_1} + c_2 \gamma_{j_2} + \cdots + c_s \gamma_{j_s} = \alpha$$

two possibly different finite F-linear combinations of elements of $\mathcal{C}$. By adding zero terms to each equation, we may assume that the vectors of $\mathcal{C}$ that occur in each equation are the same, and then by renumbering we may assume that $r = s$ and $\gamma_{i_k} = \gamma_{j_k}$. Thus, we have

$$b_1 \gamma_{i_1} + b_2 \gamma_{i_2} + \cdots + b_r \gamma_{i_r} = \alpha$$

and

$$c_1 \gamma_{i_1} + c_2 \gamma_{i_2} + \cdots + c_r \gamma_{i_r} = \alpha$$

Subtracting the second equation from the first then yields

$$(b_1 - c_1)\gamma_{i_1} + (b_2 - c_2)\gamma_{i_2} + \cdots + (b_r - c_r)\gamma_{i_r} = \alpha - \alpha = 0$$

Since $\mathcal{C}$ is linearly independent, each of these coefficients must vanish. Thus $b_i - c_i = 0$, so $b_i = c_i$ and the coefficients for α are unique. $\square$

Thus $\mathcal{B}$ is a basis if and only if it is a linearly independent spanning set. We now consider ways to find bases.

Suppose $\mathcal{C}$ is a subset of V. If $\mathcal{C}$ spans V, then obviously any set bigger than $\mathcal{C}$ also spans V. However, it is not true that we can indiscriminately remove elements from $\mathcal{C}$ and still maintain the spanning property. We say that $\mathcal{C}$ is a *minimal spanning set* of V if $\mathcal{C}$ spans V, but for every vector $\gamma \in \mathcal{C}$, the set $\mathcal{C} \setminus \{\gamma\}$ does not span.

In the other direction, suppose $\mathcal{C}$ is a linearly independent subset of V. Then clearly, every subset of $\mathcal{C}$ is also linearly independent, but we cannot add vectors indiscriminately to $\mathcal{C}$ and still maintain this property. We say that $\mathcal{C}$ is a *maximal linearly independent set* if $\mathcal{C}$ is linearly independent, but for every vector $\gamma \in V \setminus \mathcal{C}$, the set $\mathcal{C} \cup \{\gamma\}$ is not linearly independent. The interrelations between these definitions is given by

THEOREM 4.1. *Let V be a vector space over F and let $\mathcal{C}$ be a subset of V. The following are equivalent.*

   i. $\mathcal{C}$ is a basis for V.
   ii. $\mathcal{C}$ is a linearly independent spanning set.
   iii. $\mathcal{C}$ is a minimal spanning set.
   iv. $\mathcal{C}$ is a maximal linearly independent set.

PROOF. We have already observed that (i) and (ii) are equivalent. We prove that (ii) and (iii) are equivalent and then that (ii) and (iv) are equivalent.

$(ii) \Rightarrow (iii)$. Since $\mathcal{C}$ is a linearly independent spanning set, it is certainly a spanning set. Suppose $\mathcal{C}$ is not minimal. Then we can choose $\gamma \in \mathcal{C}$ such that $\mathcal{C} \setminus \{\gamma\}$ spans V. Since $\gamma \in V$, this means that we can write γ as a finite F-linear sum of elements of $\mathcal{C} \setminus \{\gamma\}$. That is, there exists $\gamma_1, \gamma_2, \ldots, \gamma_m \in \mathcal{C} \setminus \{\gamma\}$ and $a_1, a_2, \ldots, a_m \in F$ with

$$\gamma = a_1\gamma_1 + a_2\gamma_2 + \cdots + a_m\gamma_m$$

But then

$$(-1)\gamma + a_1\gamma_1 + a_2\gamma_2 + \cdots + a_m\gamma_m = 0$$

is a nontrivial F-linear sum of elements of $\mathcal{C}$ which adds to zero, and this contradicts the fact that $\mathcal{C}$ is linearly independent. Thus $\mathcal{C}$ is a minimal spanning set.

$(iii) \Rightarrow (ii)$. Since $\mathcal{C}$ is a minimal spanning set, it spans V and we need only show that it is linearly independent. If this is not the case, then there exist $\gamma_1, \gamma_2, \ldots, \gamma_m \in \mathcal{C}$ and $a_1, a_2, \ldots, a_m \in F$ not all zero, with

$$a_1\gamma_1 + a_2\gamma_2 + \cdots + a_m\gamma_m = 0$$

Since the a_i are not all zero, say $a_1 \neq 0$. Then multiplying the above equation by $-a_1^{-1}$, we see that we can assume that $a_1 = -1$. This yields

$$\gamma_1 = a_2\gamma_2 + \cdots + a_m\gamma_m \qquad (*)$$

We show now that $\mathcal{C} \setminus \{\gamma_1\}$ spans V. Let $\alpha \in V$. Since $\mathcal{C}$ spans V, there exist field elements $b_1, b_2, \ldots, b_m$ and $c_{i_1}, c_{i_2}, \ldots, c_{i_r}$ with

$$\alpha = b_1\gamma_1 + b_2\gamma_2 + \cdots + b_m\gamma_m + c_{i_1}\gamma_{i_1} + c_{i_2}\gamma_{i_2} + \cdots + c_{i_r}\gamma_{i_r}$$

where $\gamma_{i_1}, \gamma_{i_2}, \ldots, \gamma_{i_r}$ are additional elements of $\mathcal{C}$. But, by plugging in the formula $(*)$ for γ_1 in this equation, we see that

$$\alpha = b_1(a_2\gamma_2 + \cdots + a_m\gamma_m)$$
$$+ b_2\gamma_2 + b_3\gamma_3 + \cdots + b_m\gamma_m + c_{i_1}\gamma_{i_1} + c_{i_2}\gamma_{i_2} + \cdots + c_{i_r}\gamma_{i_r}$$
$$= (b_1a_2 + b_2)\gamma_2 + (b_1a_3 + b_3)\gamma_3 + \cdots + (b_1a_m + b_m)\gamma_m$$
$$+ c_{i_1}\gamma_{i_1} + c_{i_2}\gamma_{i_2} + \cdots + c_{i_r}\gamma_{i_r}$$

is an F-linear combination of elements of $\mathcal{C} \setminus \{\gamma_1\}$. Thus $\mathcal{C} \setminus \{\gamma_1\}$ spans V, and this contradicts the fact that $\mathcal{C}$ is a minimal spanning set. We therefore must have had $\mathcal{C}$ linearly independent.

$(ii) \Rightarrow (iv)$. Since $\mathcal{C}$ is a linearly independent spanning set, it is certainly independent. Let $\gamma \in V \setminus \mathcal{C}$. Since $\mathcal{C}$ spans V, there exist $\gamma_1, \gamma_2, \ldots, \gamma_m \in \mathcal{C}$ and $a_1, a_2, \ldots, a_m \in F$ with

$$\gamma = a_1 \gamma_1 + a_2 \gamma_2 + \cdots + a_m \gamma_m$$

or

$$(-1)\gamma + a_1 \gamma_1 + a_2 \gamma_2 + \cdots + a_m \gamma_m = 0$$

But this dependence relation implies that the set $\mathcal{C} \cup \{\gamma\}$ is linearly dependent. Hence $\mathcal{C}$ is a maximal linearly independent set.

$(iv) \Rightarrow (ii)$. Finally we are given that $\mathcal{C}$ is a maximal linearly independent set, so we need only show that it spans. Let $\alpha \in V$. If $\alpha \in \mathcal{C}$, then

$$\alpha = (1)\alpha$$

shows that α is an F-linear sum of elements of $\mathcal{C}$. Now let $\alpha \notin \mathcal{C}$ so that $\mathcal{C} \cup \{\alpha\}$ is properly bigger than $\mathcal{C}$. By the maximality property of $\mathcal{C}$, we know that this bigger set is linearly dependent, so there exist $\gamma_1, \gamma_2, \ldots, \gamma_m \in \mathcal{C}$ and $a, a_1, a_2, \ldots, a_m \in F$ not all zero with

$$a\alpha + a_1 \gamma_1 + a_2 \gamma_2 + \cdots + a_m \gamma_m = 0$$

Can $a = 0$? If this were the case, then we could delete the α term above and have a nontrivial F-linear sum of elements of $\mathcal{C}$ adding to zero, a contradiction since $\mathcal{C}$ is linearly independent. Thus $a \neq 0$ and multiplying the above equation by $-a^{-1}$, we see that we can assume that $a = -1$. Thus we have

$$(-1)\alpha + a_1 \gamma_1 + a_2 \gamma_2 + \cdots + a_m \gamma_m = 0$$

so

$$\alpha = a_1 \gamma_1 + a_2 \gamma_2 + \cdots + a_m \gamma_m$$

and $\mathcal{C}$ spans V. This completes the proof. $\square$

Finally, one more definition. Let V be a vector space over F. If V has a finite spanning set, then we say that V is *finite dimensional*. Otherwise, V is *infinite dimensional*. It is apparent that finite dimensional spaces are in some sense small and therefore this property should carry over to subspaces. However this fact requires proof and we will prove it and more in the next section. We close this section with

COROLLARY 4.1. *Let V be a vector space with a finite spanning set $\mathcal{C}$. Then some subset $\mathcal{B}$ of $\mathcal{C}$ is a basis. In particular, finite dimensional vector spaces have bases.*

PROOF. $\mathcal{C}$ has only finitely many subsets and one of these subsets, namely $\mathcal{C}$ itself spans V. Thus we may choose a subset $\mathcal{B}$ of $\mathcal{C}$ that is a spanning set of smallest possible size. Clearly $\mathcal{B}$ is a minimal spanning set of V and therefore by Theorem 4.1, $\mathcal{B}$ is a basis.

If V is a finite dimensional vector space, then V does have such a finite spanning set, and hence V does have a basis. $\square$

It is in fact true that every vector space has a basis. However a proof of this requires going beyond the above finite type argument, and one must use transfinite methods. We will put this study off until the end of these notes since, for the most part, we are concerned with finite dimensional spaces.

Problems

4.1. Find a basis for the solution space of the real linear equation

$$x_1 - 2x_2 + x_3 - x_4 = 0$$

4.2. Find a basis for the space of simultaneous solutions of the real linear equations

$$x_1 - 2x_2 + x_3 - x_4 = 0$$
$$2x_1 - 3x_2 - x_3 + 2x_4 = 0$$

4.3. For each $i \geq 0$, let $\beta_i \in F[x]$ be a polynomial of degree i. Prove that $\mathcal{B} = \{\beta_0, \beta_1, \beta_2, \ldots\}$ is a basis for $F[x]$.

4.4. Verify that $\mathcal{B} = \{\beta_1, \beta_2, \ldots, \beta_n\}$ as given in Example 4.3 is a basis for F^n.

4.5. Let V be a vector space with basis $\mathcal{B} = \{\beta_1, \beta_2, \beta_3, \beta_4\}$. Prove that $\mathcal{C} = \{\beta_1, \beta_2 - \beta_1, \beta_3 - \beta_2, \beta_4 - \beta_3\}$ is also a basis.

4.6. Prove that the subset

$$\mathcal{C} = \{(1, 0, -1, 2), (1, 1, 0, 1), (1, 0, 2, 3), (0, 0, 1, 2)\}$$

of $\mathbb{R}^4$ is linearly independent.

4.7. Prove that the subset $\mathcal{C} = \{(1, 0, 2), (1, 1, 0), (1, 0, -1)\}$ of $\mathbb{Q}^3$ is a spanning set.

4.8. Let $\gamma \in V$. Show that $\{\gamma\}$ is linearly independent if and only if $\gamma \neq 0$.

4.9. Let $\mathcal{C}$ be a subset of vector space V. Show that $\mathcal{C}$ is linearly independent if and only if $\mathcal{C}$ is a basis for some subspace of V.

4.10. Let $F \subseteq L \subseteq K$ be a chain of three fields. Suppose $\mathcal{A} = \{\alpha_1, \alpha_2, \ldots, \alpha_n\}$ is a basis for L as a vector space over F and suppose $\mathcal{B} = \{\beta_1, \beta_2, \ldots, \beta_m\}$ is a basis for K as a vector space over L. Prove that

$$\{\alpha_i \beta_j \mid i = 1, 2, \ldots, n; \quad j = 1, 2, \ldots, m\}$$

is a basis for K as a vector space over F.

5.   The Replacement Theorem

In this section, we prove our first real theorem. It is quite elementary, but never-the-less it tells us almost everything of a theoretical nature that we might want to know about finite dimensional vector spaces.

THEOREM 5.1 (Replacement Theorem). *Let V be a vector space over the field F, let $\mathcal{A} = \{\alpha_1, \alpha_2, \ldots, \alpha_n\}$ be a linearly independent subset of V, and let $\mathcal{C} = \{\gamma_1, \gamma_2, \ldots, \gamma_m\}$ be a spanning set of V. Then by suitably renumbering the elements of $\mathcal{C}$, if necessary, it follows that the set*

$$\{\alpha_1, \alpha_2, \ldots, \alpha_n, \gamma_{n+1}, \ldots, \gamma_m\}$$

spans V. In particular, $m \geq n$.

In other words, we can replace γ_1 by α_1, γ_2 by α_2, $\ldots$ and γ_n by α_n in $\mathcal{C}$, and the set will still span V. Observe that the renumbering of $\mathcal{C}$ is really necessary. First, the order in which the γ's are written is of course accidental, and there is no reason to believe a priori that the given order might be consistent with that of the α's to allow the replacement to work. Second, consider the extreme case where $\mathcal{C}$ is a basis and $\mathcal{A} \subseteq \mathcal{C}$. Then $\mathcal{C}$ is a minimal spanning set and from this it is clear that the only possible replacement that can work for all n requires $\alpha_1 = \gamma_1$, $\alpha_2 = \gamma_2$, $\ldots$, $\alpha_n = \gamma_n$.

PROOF. We show by induction on j, for $0 \leq j \leq n$, that we can choose elements, which we label $\gamma_1, \gamma_2, \ldots, \gamma_j \in \mathcal{C}$ so that the set

$$\mathcal{C}_j = \{\alpha_1, \alpha_2, \ldots, \alpha_j, \gamma_{j+1}, \gamma_{j+2}, \ldots, \gamma_m\}$$

spans V.

Observe that the induction starts at $j = 0$. Here no replacements are made, so by assumption, $\mathcal{C}_0 = \mathcal{C}$ spans V.

Now let us assume that the first j γ's have been so labeled and that the set $\mathcal{C}_j$ above spans V. If $j = n$, we are done. If not then $j < n$ so $\alpha_{j+1} \in V$. In particular, since $\mathcal{C}_j$ spans V, there exist field elements $a_1, a_2, \ldots, a_j, b_{j+1}, b_{j+2}, \ldots, b_m$ with

$$\alpha_{j+1} = a_1\alpha_1 + a_2\alpha_2 + \cdots + a_j\alpha_j$$
$$+ b_{j+1}\gamma_{j+1} + b_{j+2}\gamma_{j+2} + \cdots + b_m\gamma_m \qquad (*)$$

Can all the b_i's be zero? If this were the case, then

$$\alpha_{j+1} = a_1\alpha_1 + a_2\alpha_2 + \cdots + a_j\alpha_j$$

so

$$0 = a_1\alpha_1 + a_2\alpha_2 + \cdots + a_j\alpha_j + (-1)\alpha_{j+1} + 0\alpha_{j+2} + \cdots + 0\alpha_n$$

is a nontrivial F-linear sum of the elements of $\mathcal{A}$ that adds to zero. But $\mathcal{A}$ is linearly independent, so this cannot occur. Thus some b_i is nonzero.

By relabeling the elements $\gamma_{j+1}, \gamma_{j+2}, \ldots, \gamma_m$ if necessary, we can assume that $b_{j+1} \neq 0$. With this assumption, we show that

$$\mathcal{C}_{j+1} = \{\alpha_1, \alpha_2, \ldots, \alpha_j, \alpha_{j+1}, \gamma_{j+2}, \gamma_{j+3}, \ldots, \gamma_m\}$$

spans V. First, since $b_{j+1} \neq 0$, we can solve for γ_{j+1} in (∗) and obtain

$$\gamma_{j+1} = \bar{a}_1 \alpha_1 + \bar{a}_2 \alpha_2 + \cdots + \bar{a}_j \alpha_j + \bar{a}_{j+1} \alpha_{j+1}$$
$$+ \bar{b}_{j+2} \gamma_{j+2} + \bar{b}_{j+3} \gamma_{j+3} + \cdots + \bar{b}_m \gamma_m \qquad (\ast\ast)$$

where

$$\bar{a}_i = -b_{j+1}^{-1} a_i \qquad\qquad \text{for } i \leq j$$
$$\bar{a}_{j+1} = b_{j+1}^{-1}$$
$$\bar{b}_i = -b_{j+1}^{-1} b_i \qquad\qquad \text{for } i \geq j+2$$

Now let $\beta \in V$ be arbitrary. Since $\mathcal{C}_j$ spans V, there exist field elements $c_1, c_2, \ldots, c_j, d_{j+1}, d_{j+2}, \ldots, d_m$ with

$$\beta = c_1 \alpha_1 + c_2 \alpha_2 + \cdots + c_j \alpha_j$$
$$+ d_{j+1} \gamma_{j+1} + d_{j+2} \gamma_{j+2} + \cdots + d_m \gamma_m$$

Plugging the formula (∗∗) for γ_{j+1} into this equation, we get easily

$$\beta = c_1 \alpha_1 + c_2 \alpha_2 + \cdots + c_j \alpha_j$$
$$+ d_{j+1}(\bar{a}_1 \alpha_1 + \bar{a}_2 \alpha_2 + \cdots + \bar{a}_j \alpha_j + \bar{a}_{j+1} \alpha_{j+1})$$
$$+ d_{j+1}(\bar{b}_{j+2} \gamma_{j+2} + \bar{b}_{j+3} \gamma_{j+3} + \cdots + \bar{b}_m \gamma_m)$$
$$+ d_{j+2} \gamma_{j+2} + \cdots + d_m \gamma_m$$
$$= \bar{c}_1 \alpha_1 + \bar{c}_2 \alpha_2 + \cdots + \bar{c}_j \alpha_j + \bar{c}_{j+1} \alpha_{j+1}$$
$$+ \bar{d}_{j+2} \gamma_{j+2} + \bar{d}_{j+3} \gamma_{j+3} + \cdots + \bar{d}_m \gamma_m$$

where

$$\bar{c}_i = c_i + d_{j+1} \bar{a}_i \qquad\qquad \text{for } i \leq j$$
$$\bar{c}_{j+1} = d_{j+1} \bar{a}_{j+1}$$
$$\bar{d}_i = d_i + d_{j+1} \bar{b}_i \qquad\qquad \text{for } i \geq j+2$$

Thus β is written as an F-linear sum of elements of $\mathcal{C}_{j+1}$. Since this is true for all such $\beta \in V$, this says that $\mathcal{C}_{j+1}$ spans V.

Therefore the induction step is verified. The first part of the theorem now follows from the case $j = n$. Finally, since n different γ's have been replaced, we must clearly have $m \geq n$. This completes the proof. $\square$

To some readers, the proof of the last statement $m \geq n$ may be a little disturbing. Where in the proof do we really show that we do not run out of elements of $\mathcal{C}$ in this process? The answer is precisely in studying equation $(*)$. If there were no γ's left at this stage, then certainly all the b_i's would be zero. Thus, when we show that some b_i is not zero, we are in fact also showing that some γ_i exists with $i > j$.

The Replacement Theorem has numerous corollaries. We start with the most important. If $\mathcal{A}$ is a subset of V, we let $|\mathcal{A}|$ denote its size.

THEOREM 5.2. *Let V be a finite dimensional vector space over F. Then there exists a unique integer $n \geq 0$ called the dimension of V with the property that*

    i. If $\mathcal{A}$ is a linearly independent subset of V, then $|\mathcal{A}| \leq n$.
    ii. If $\mathcal{B}$ is a basis of V, then $|\mathcal{B}| = n$.
    iii. If $\mathcal{C}$ is a spanning subset of V, then $|\mathcal{C}| \geq n$.

PROOF. Since V is a finite dimensional vector space, we know that it has a finite basis $\widetilde{\mathcal{B}}$. Fix such a basis and let $n = |\widetilde{\mathcal{B}}|$. We show that n satisfies (i), (ii) and (iii). Let $\mathcal{A}$, $\mathcal{B}$ and $\mathcal{C}$ be as above.

Let $\mathcal{A}_0$ be a finite subset of $\mathcal{A}$. Then $\mathcal{A}_0$ is linearly independent and $\widetilde{\mathcal{B}}$ spans V. Thus by the Replacement Theorem, $|\mathcal{A}_0| \leq |\widetilde{\mathcal{B}}| - n$. Since $|\mathcal{A}_0| \leq n$ for all such finite subsets $\mathcal{A}_0$ of $\mathcal{A}$, this clearly implies that $\mathcal{A}$ is finite and $|\mathcal{A}| \leq n$. Thus (i) is proved.

If $|\mathcal{C}| = \infty$, then certainly $|\mathcal{C}| \geq n$. Thus, in proving (iii), we can assume that $\mathcal{C}$ is finite. Now $\widetilde{\mathcal{B}}$ is linearly independent and $\mathcal{C}$ spans V. Hence by the Replacement Theorem, $|\mathcal{C}| \geq |\widetilde{\mathcal{B}}| = n$ and (iii) follows.

Finally, $\mathcal{B}$ is linearly independent, so by (i) with $\mathcal{A} = \mathcal{B}$ we have $|\mathcal{B}| \leq n$. Also $\mathcal{B}$ spans, so by (iii) with $\mathcal{C} = \mathcal{B}$ we have $|\mathcal{B}| \geq n$. This yields $|\mathcal{B}| = n$ and (ii) follows. Note, by (ii), n is the common size of all bases of V, and thus clearly n is uniquely determined. $\square$

We use $\dim_F V$ to denote the dimension of V.

COROLLARY 5.1. *Let $\dim_F V = n < \infty$ and let $\mathcal{A}$ be a subset of V of size n. Then the following are equivalent.*

    i. $\mathcal{A}$ is linearly independent.
    ii. $\mathcal{A}$ is a basis.
    iii. $\mathcal{A}$ spans V.

PROOF. Certainly (ii) implies (i) and (iii). Now assume that $\mathcal{A}$ satisfies (i). Then $\mathcal{A}$ is linearly independent and $|\mathcal{A}| = n$. But there can be no linearly independent subsets of size $n + 1$, so $\mathcal{A}$ is a maximal linearly independent subset and hence is a basis for V. Finally, let $\mathcal{A}$ satisfy (iii). Then $\mathcal{A}$ spans V and $|\mathcal{A}| = n$. But there can be no spanning set of V of size $n - 1$, so $\mathcal{A}$ is a minimal spanning set and hence it is a basis for V. $\square$

At this point, a natural question to consider is whether a subspace of a finite dimensional vector space must also be finite dimensional. We proceed to answer this now.

LEMMA 5.1. *Let V be a finite dimensional vector space and let $\mathcal{A}$ be a linearly independent subset of V. Then $\mathcal{A}$ can be extended to a basis of V.*

PROOF. Let $\mathcal{A} = \{\alpha_1, \alpha_2, \ldots, \alpha_r\}$ and let $\mathcal{B} = \{\beta_1, \beta_2, \ldots, \beta_n\}$ be a basis for V of size $n = \dim_F V$. Since $\mathcal{A}$ is linearly independent and $\mathcal{B}$ spans V, the Replacement Theorem implies that for a suitable ordering of the β's, the set

$$\widetilde{\mathcal{A}} = \{\alpha_1, \alpha_2, \ldots, \alpha_r, \beta_{r+1}, \beta_{r+2}, \ldots, \beta_n\}$$

spans V. Clearly $|\widetilde{\mathcal{A}}| \leq n$ (since some duplication might occur), but $\widetilde{\mathcal{A}}$ spans V, so we must have $|\widetilde{\mathcal{A}}| \geq n$. Thus $\widetilde{\mathcal{A}}$ is a spanning set of size n and, by the above corollary, $\widetilde{\mathcal{A}}$ is a basis for V. $\square$

THEOREM 5.3. *Let W be a subspace of a finite dimensional vector space V. Then*

    i. W is finite dimensional and $\dim_F W \leq \dim_F V$.
    ii. $\dim_F W = \dim_F V$ if and only if $W = V$.
    iii. Any basis of W can be extended to a basis for V.
    iv. There exists a subspace W' of V such that $V = W \oplus W'$.

PROOF. Say $\dim_F V = n$. Let $\mathcal{A}$ be a linearly independent subset of W. Then no nontrivial finite F-linear sum of elements of $\mathcal{A}$ can add to zero. Furthermore, the arithmetic in W is the same as in V, so the previous statement holds whether we view $\mathcal{A}$ as a subset of W or of V. In particular, $\mathcal{A}$ is linearly independent in V. By the definition of the dimension of V, we have $|\mathcal{A}| \leq n$.

We start with $\mathcal{A}_0 = \emptyset$, the empty subset of W. Then $\mathcal{A}_0$ is linearly independent in W. If it is maximal with this property, then $\mathcal{A}_0$ is a basis of W, and of course $W = 0$. If $\mathcal{A}_0$ is not maximal, then we can adjoin a suitable vector of W to get $\mathcal{A}_1$, a linearly independent subset of W of size 1. If $\mathcal{A}_1$ is not maximal, we can adjoin a suitable vector of W to get $\mathcal{A}_2$, a linearly independent subset of W of size 2. We continue with this process and, by the argument of the previous paragraph, it must terminate at some $\mathcal{A}_j$ with $j \leq n$. Thus $\mathcal{A}_j$ is a maximal linearly independent subset of W, so $\mathcal{A}_j$ is a basis for W. This shows that W has a finite spanning set, so it is finite dimensional. Then, by the definition of dimension, we have $\dim_F W = j \leq n$. This yields (i).

Now let $\mathcal{A} = \{\alpha_1, \alpha_2, \ldots, \alpha_j\}$ be any basis for W. Since $\mathcal{A}$ is linearly independent in V, the above lemma implies that we can extend $\mathcal{A}$ to $\mathcal{B} = \{\alpha_1, \alpha_2, \ldots, \alpha_j, \beta_{j+1}, \beta_{j+2}, \ldots, \beta_n\}$, a basis for V, thereby proving (iii).

Now if $W = V$, then certainly $\dim_F W = \dim_F V$. Conversely, suppose $\dim_F W = \dim_F V$. Then we must have $\mathcal{B} = \mathcal{A}$ above and in particular no β's occur. If $\gamma \in V$, then γ can be written as an F-linear sum of elements of $\mathcal{B} = \mathcal{A}$. Since $\mathcal{A} \subseteq W$, this implies that $\gamma \in W$ and thus $V \subseteq W$. Therefore clearly $V = W$ and (ii) follows.

Finally, set $W' = \langle \beta_{j+1}, \beta_{j+2}, \ldots, \beta_n \rangle$. We show that $V = W \oplus W'$. First, let $\gamma \in V$. Since $\mathcal{B}$ spans V, we have

$$\gamma = a_1\alpha_1 + a_2\alpha_2 + \cdots + a_j\alpha_j + b_{j+1}\beta_{j+1} + b_{j+2}\beta_{j+2} + \cdots + b_n\beta_n$$

for suitable elements $a_1, a_2, \ldots, a_j, b_{j+1}, b_{j+2}, \ldots, b_n \in F$. If we set $\alpha = a_1\alpha_1 + a_2\alpha_2 + \cdots + a_j\alpha_j$ and $\beta = b_{j+1}\beta_{j+1} + b_{j+2}\beta_{j+2} + \cdots + b_n\beta_n$, then clearly $\alpha \in W$, $\beta \in W'$ and $\gamma = \alpha + \beta \in W + W'$. Thus $V = W + W'$.

Now suppose $\delta \in W \cap W'$. Then from $\delta \in W$ and $\delta \in W'$, we deduce that

$$\delta = c_1\alpha_1 + c_2\alpha_2 + \cdots + c_j\alpha_j$$
$$\delta = d_{j+1}\beta_{j+1} + d_{j+2}\beta_{j+2} + \cdots + d_n\beta_n$$

for suitable $c_1, c_2, \ldots, c_j, d_{j+1}, d_{j+2}, \ldots, d_n \in F$. Thus

$$c_1\alpha_1 + c_2\alpha_2 + \cdots + c_j\alpha_j$$
$$+ (-d_{j+1})\beta_{j+1} + (-d_{j+2})\beta_{j+2} + \cdots + (-d_n)\beta_n = 0$$

Since $\mathcal{B} = \{\alpha_1, \alpha_2, \ldots, \alpha_j, \beta_{j+1}, \beta_{j+2}, \ldots, \beta_n\}$ is linearly independent, we conclude that all coefficients above must vanish. In particular, $c_1 = c_2 = \cdots = c_j = 0$, so $\delta = 0$ and $W \cap W' = 0$. Thus $V = W + W' = W \oplus W'$ and the theorem is proved. $\square$

We remark that W' above is called a *complement* for W in V. Of course, W' is by no means unique. We now consider a general relation between two subspaces of V.

THEOREM 5.4. *Let W_1 and W_2 be two subspaces of a finite dimensional vector space V. Then*

$$\dim_F(W_1 + W_2) + \dim_F(W_1 \cap W_2) = \dim_F W_1 + \dim_F W_2$$

PROOF. We know that all of the above spaces are finite dimensional. Let $\mathcal{C} = \{\gamma_1, \gamma_2, \ldots, \gamma_t\}$ be a basis for $W_1 \cap W_2$, so that $\dim_F(W_1 \cap W_2) = t$. Since $W_1 \cap W_2$ is a subspace of W_1, we can extend $\mathcal{C}$ to a basis $\mathcal{A} = \{\gamma_1, \gamma_2, \ldots, \gamma_t, \alpha_1, \alpha_2, \ldots, \alpha_r\}$ of W_1. Similarly, we can extend $\mathcal{C}$ to a basis $\mathcal{B} = \{\gamma_1, \gamma_2, \ldots, \gamma_t, \beta_1, \beta_2, \ldots, \beta_s\}$ of W_2. Observe that $\dim_F W_1 = t + r$ and $\dim_F W_2 = t + s$, so

$$\dim_F W_1 + \dim_F W_2 - \dim_F(W_1 \cap W_2)$$
$$= (t + r) + (t + s) - t = r + s + t$$

In other words, what we want to prove is that $\dim_F(W_1 + W_2) = r + s + t$. Now we have a nice set

$$\mathcal{D} = \{\alpha_1, \alpha_2, \ldots, \alpha_r, \beta_1, \beta_2, \ldots, \beta_s, \gamma_1, \gamma_2, \ldots, \gamma_t\}$$

of $r + s + t$ seemingly distinct vectors that are all clearly contained in $W_1 + W_2$. Thus the obvious approach is to prove that $\mathcal{D}$ is a basis for $W_1 + W_2$.

First let $\delta \in W_1 + W_2$. Then $\delta = \alpha + \beta$ with $\alpha \in W_1$ and $\beta \in W_2$. Since $\mathcal{A}$ spans W_1 and $\mathcal{B}$ spans W_2, we have

$$\alpha = a_1\alpha_1 + \cdots + a_r\alpha_r + c_1\gamma_1 + \cdots + c_t\gamma_t$$
$$\beta = b_1\beta_1 + \cdots + b_s\beta_s + d_1\gamma_1 + \cdots + d_t\gamma_t$$

for suitable field elements a_i, b_i, c_i and d_i. Thus

$$\delta = \alpha + \beta = a_1\alpha_1 + \cdots + a_r\alpha_r + b_1\beta_1 + \cdots + b_s\beta_s$$
$$+ (c_1 + d_1)\gamma_1 + \cdots + (c_t + d_t)\gamma_t$$

and $\mathcal{D}$ spans $W_1 + W_2$.

Now suppose that some F-linear sum of the elements of $\mathcal{D}$ sums to zero. Say

$$a_1\alpha_1 + \cdots + a_r\alpha_r + b_1\beta_1 + \cdots + b_s\beta_s + c_1\gamma_1 + \cdots + c_t\gamma_t = 0$$

Then

$$a_1\alpha_1 + \cdots + a_r\alpha_r + c_1\gamma_1 + \cdots + c_t\gamma_t = (-b_1)\beta_1 + \cdots + (-b_s)\beta_s$$

Now the above left-hand side is clearly in W_1 and the right-hand side is in W_2, so this common vector which we call δ is in $W_1 \cap W_2$. Thus, since $\mathcal{C}$ spans $W_1 \cap W_2$, we have

$$\delta = d_1\gamma_1 + \cdots + d_t\gamma_t$$

for suitable $d_i \in F$. This yields the equations

$$a_1\alpha_1 + \cdots + a_r\alpha_r + c_1\gamma_1 + \cdots + c_t\gamma_t = d_1\gamma_1 + \cdots + d_t\gamma_t$$
$$(-b_1)\beta_1 + \cdots + (-b_s)\beta_s = d_1\gamma_1 + \cdots + d_t\gamma_t$$

or equivalently

$$a_1\alpha_1 + \cdots + a_r\alpha_r + (c_1 - d_1)\gamma_1 + \cdots + (c_t - d_t)\gamma_t = 0$$
$$b_1\beta_1 + \cdots + b_s\beta_s + d_1\gamma_1 + \cdots + d_t\gamma_t = 0$$

Therefore, by the linear independence of $\mathcal{A}$ and $\mathcal{B}$, we conclude that

$$a_1 = \cdots = a_r = 0, \qquad b_1 = \cdots = b_s = 0, \qquad d_1 = \cdots = d_t = 0$$

and then $c_1 = \cdots = c_t = 0$. Thus $\mathcal{D}$ is linearly independent and hence a basis for $W_1 + W_2$. Observe that the above also tells us that all $r+s+t$ elements of $\mathcal{D}$ are distinct so

$$\dim_F(W_1 + W_2) = |\mathcal{D}| = r + s + t$$

and the result follows. □

Problems

5.1. Suppose that A and B are finite disjoint index sets. Convince yourself that for vectors $\alpha_i \in V$ we have

$$\sum_{i \in A \cup B} \alpha_i = \sum_{i \in A} \alpha_i + \sum_{i \in B} \alpha_i$$

What does this say when $A = \emptyset$ is the empty set? What is a basis for the space $V = 0$?

Find a basis and the dimension of each of the following F-vector spaces.

5.2. $V = F^n$.

5.3. $V = F[x]$.

5.4. $V = \mathbb{C}$ over the field $F = \mathbb{R}$.

5.5. V is the set a functions from a finite set U into the field F.

Let V be a vector space of finite dimension n.

5.6. Let W_1 and W_2 be subspaces of V with

$$\dim_F W_1 + \dim_F W_2 > n$$

Prove that $W_1 \cap W_2 \neq 0$.

5.7. Let $V = W_k > W_{k-1} > \cdots > W_1 > W_0 = 0$ be a finite chain of distinct subspaces of V. Show that $k \leq n$.

5.8. Consider the chain of subspaces in the preceding problem. Show that there exist subspaces $U_0, U_1, \ldots, U_k$ of V such that

$$W_i = U_0 \oplus U_1 \oplus \cdots \oplus U_i$$

for all i.

5.9. Find two different complements for the subspace W of V where $W = \langle (1, -1, 2, 0), (1, 1, 1, -1) \rangle$ and $V = \mathbb{Q}^4$.

5.10. Let $F \subseteq L \subseteq K$ be a chain of fields. Prove that

$$\dim_F K = (\dim_F L)(\dim_L K)$$

in the sense that if any two of these three numbers is finite, then so is the third and the product formula holds. (See Problem 4.10.)

6.   Matrices and Elementary Operations

Now we introduce a bit of notation and some computational techniques. We are concerned with the following two rather concrete problems. Given a finite collection $\mathcal{A} = \{\alpha_1, \alpha_2, \ldots, \alpha_m\}$ of vectors in F^n, find a basis for the subspace $V = \langle \mathcal{A} \rangle$ spanned by $\mathcal{A}$. Of course, canonical bases do not exist in general, so we could just try to find a nice basis or we could look for a basis that is a subset of $\mathcal{A}$.

To start with, let us write

$$\alpha_i = (a_{i1}, a_{i2}, \ldots, a_{in}) \in F^n$$

for $i = 1, 2, \ldots, m$. Since the entries are double subscripted, it is tempting to put them into a rectangular array which we call a *matrix*. We "hold these entries in place" by surrounding the array A with square brackets, but we no longer use commas to separate the entries since this would not work in the vertical direction. Thus we write

$$A = \begin{bmatrix} a_{11} & a_{12} & \ldots & a_{1n} \\ a_{21} & a_{22} & \ldots & a_{2n} \\ & \cdots \cdots \cdots & \\ a_{m1} & a_{m2} & \ldots & a_{mn} \end{bmatrix}$$

for the corresponding $m \times n$ (m by n) matrix. As usual the double subscripts should be separated by a comma, but we will ignore this if there is no possibility of confusion. We frequently abbreviate matrix notation by writing $A = [a_{ij}]$, and we denote the set of all $m \times n$ matrices by $F^{m \times n}$.

This matrix has m rows, labeled 1 through m going down, and n columns, labeled 1 through n from left to right. Thus a_{ij} is the entry in the ith row and jth column. Now the rows each have n ordered entries and hence can be viewed as elements of F^n. Indeed these correspond to the original vectors $\mathcal{A} = \{\alpha_1, \alpha_2, \ldots, \alpha_m\}$. The subspace of F^n spanned by these *row vectors* is called the *row space* of A. Similarly, we note that the columns of A each have m ordered entries. Thus by mentally rotating these columns $90°$, we can view these as vectors in F^m. The subspace of F^m spanned by these *column vectors* is then called the *column space* of A.

Now to find a basis for $V = \langle \mathcal{A} \rangle$ we are free to modify the vectors in $\mathcal{A}$, and this gives rise to corresponding operations on the rows of the matrix. Indeed, we list the three so-called *elementary row operations* that are of interest.

R1. Interchange any pair of rows. More precisely, if we interchange rows i and k, we write this operation as $\mathfrak{R}_1(i, k)$.

R2. Multiply the ith row by a nonzero constant c so that the entry a_{ij} becomes $c \cdot a_{ij}$ for all $j = 1, 2, \ldots, n$. We denote this operation by $\mathfrak{R}_2(i; c)$.

R3. Finally, for $i \neq k$, we add $c \in F$ times the kth row to the ith, so that the entry a_{ij} becomes $a_{ij} + c \cdot a_{kj}$ for all $j = 1, 2, \ldots, n$. We denote this operation by $\mathfrak{R}_3(i, k; c)$.

Notice that $\mathfrak{R}_2(i; 1)$ and $\mathfrak{R}_3(i, k; 0)$ are both equal to Id, the identity operation that leaves A unchanged. The key property of all these operations is that they are invertible. Indeed, one can undo each of these with another elementary row operation.

LEMMA 6.1. *Each elementary row operation is invertible with its inverse being an elementary row operation of the same type.*

PROOF. Clearly $\mathfrak{R}_1(i, k)\mathfrak{R}_1(i, k) = \text{Id}$ and $\mathfrak{R}_2(i; c^{-1})\mathfrak{R}_2(i; c) = \text{Id}$. Finally since $\mathfrak{R}_3(i, k; c)$ does not change the kth row of A, we see that $\mathfrak{R}_3(i, k; -c)\mathfrak{R}_3(i, k; c) = \text{Id}$.   □

With this, we can quickly prove that elementary row operations do not change the row space. Specifically, we have

LEMMA 6.2. *If A is an $m \times n$ matrix over F, and if $\mathfrak{R}$ is an elementary row operation, then the row spaces of A and of $\mathfrak{R}(A)$ are the same.*

PROOF. Write $B = \mathfrak{R}(A)$. By considering the three operations in turn, we see easily that each row vector of B belongs to the row space of A and hence the row space of B is contained in the row space of A. Furthermore, since $\mathfrak{R}^{-1}(B) = \mathfrak{R}^{-1}(\mathfrak{R}(A)) = A$, we obtain the reverse inclusion and consequently the two row spaces are equal.   □

Now let us see what we can do with these operations. To start with, we describe a fairly nice matrix structure. We say that a matrix A is in *row echelon form* if

E1. The zero rows of A, if any, are at the bottom, that is they are all below the nonzero rows.

E2. Each nonzero row, if any, starts with a *leading* 1. That is, the first nonzero entry, from left to right, is a 1.

E3. The leading 1s slope down and to the right. Specifically, if the rows i and $i + 1$ are both nonzero, then the leading 1 in row $i+1$ is contained in a column strictly to the right of the leading 1 of row i.

For example, we have

EXAMPLE 6.1. The 5×6 real matrix

$$A = [a_{ij}] = \begin{bmatrix} 0 & 1 & 2 & 3 & 4 & 6 \\ 0 & 0 & 0 & 1 & 3 & 1 \\ 0 & 0 & 0 & 0 & 1 & 4 \\ 0 & 0 & 0 & 0 & 0 & 0 \\ 0 & 0 & 0 & 0 & 0 & 0 \end{bmatrix}$$

is in row echelon form since (1) the two zero rows are at the bottom, (2) each of the nonzero rows starts with an entry 1, and (3) the leading 1s slant down and to the right. Note that the leading 1s are the entries a_{12}, a_{24} and a_{35}. On the other hand, $a_{26} = 1$ but it is not a leading 1.

It follows from (E1) and (E3) that the entries below each leading 1 are all 0. But what about the entries above the leading 1s? We say that the matrix A is in *reduced row echelon form* if it satisfies (E1), (E2) and (E3), so that it is in row echelon form, and also satisfies

E4. In each column determined by a leading 1, all the remaining entries are 0.

EXAMPLE 6.2. The matrix A of Example 6.1 is not in reduced row echelon form, but the modified matrix

$$B = [b_{ij}] = \begin{bmatrix} 0 & 1 & 2 & 0 & 0 & 6 \\ 0 & 0 & 0 & 1 & 0 & 1 \\ 0 & 0 & 0 & 0 & 1 & 4 \\ 0 & 0 & 0 & 0 & 0 & 0 \\ 0 & 0 & 0 & 0 & 0 & 0 \end{bmatrix}$$

is in reduced row echelon form since the entries in the second, fourth and fifth columns, that are not the leading 1s, are all equal to 0.

Reduced row echelon form matrices are important because of the uniqueness property described below.

THEOREM 6.1. *Let $A \in F^{m \times n}$. Then we have*

i. Using a sequence of elementary row operations, the matrix A can be transformed to a reduced row echelon matrix A'.

ii. The nonzero rows of A' form a basis for the row space of A. In particular, the dimension of the row space of A is equal to the number of nonzero rows of A'.

iii. The matrix A' is uniquely determined by A, independent of the sequence of elementary row operations that are used.

PROOF. We show how to transform $A = [a_{ij}]$ into a reduced row echelon matrix. If all entries of A are 0, we are done, so suppose not and say the sth column is the left-most nonzero column. If $a_{rs} \neq 0$, then by interchanging the rth and first rows, if necessary, we can place a_{rs} in the $1, s$-position. Thus we may assume that $a_{1s} \neq 0$ and by multiplying the first row by a_{1s}^{-1}, we can now assume that $a_{1s} = 1$. Next, we successively add $-a_{is}$ times the first row to the ith for $i = 2, 3, \ldots, m$. This leaves the first row unchanged, but modifies the others so that all remaining entries in the sth column are 0.

Next, we consider the submatrix consisting of rows 2 through m. Note that in this matrix the first s columns are 0. Again, if all entries are 0, we are done. If not let t be the left-most nonzero column, so that $t > s$. If $a_{rt} \neq 0$, then interchanging the second and rth rows, if necessary, we can put a_{rt} in the $2, t$-position. Thus we may assume that $a_{2t} \neq 0$ and by multiplying the second row by a_{2t}^{-1}, we can now assume that $a_{2t} = 1$. Next, we successively add $-a_{it}$ times the second row to the ith for all $i = 1, 3, \ldots, m$. This leaves the sth column unchanged since $a_{2s} = 0$, but modifies the tth column so that all entries, including the first, but not the leading 1, have become 0. We now move on to the submatrix consisting of rows 3 through m and, by continuing in this manner, we can clearly transform A into a reduced row echelon matrix. With this, part (i) follows.

For parts (ii) and (iii), let $\alpha_1', \alpha_2', \ldots, \alpha_r'$ be the nonzero row vectors of A', in their natural order, with leading 1s in columns $s_1, s_2, \ldots, s_r$ respectively. If

$$\beta = x_1\alpha_1' + x_2\alpha_2' + \cdots + x_r\alpha_r'$$

with $x_1, x_2, \ldots, x_r \in F$, then the s_jth entry of β is precisely equal to x_j. In particular, if $\beta = 0$, then all $x_j = 0$ and we conclude that $\alpha_1', \alpha_2', \ldots, \alpha_r'$ are linearly independent. On the other hand, we know that the row space V of A is equal to the row space of A', so $\alpha_1', \alpha_2', \ldots, \alpha_r'$ span V. Thus $\mathcal{A} = \{\alpha_1', \alpha_2', \ldots, \alpha_r'\}$ is indeed a basis for V. This proves (ii) and consequently $r = \dim_F V$.

Finally, observe that β above is a typical element of V and our comments concerning the s_jth entry of β imply that α_r' is the unique vector in V with a leading 1 and the largest number of preceding 0s. Thus V, and hence A, determines the row vector α_r' and also the column s_r. Next consider the nonzero vectors in V with 0 entry in position s_r. If β above is such an element, then $x_r = 0$ and it follows that α_{r-1}' is the unique such vector with a leading 1 and with the largest number of preceding 0s. Thus V, and hence A, determines the row vector α_{r-1}' and also the column s_{r-1}. Continuing in this manner with vectors

$\alpha'_{r-2}, \ldots, \alpha'_1$, in turn, we see that V, and hence A, determines all the rows of A' in order. Thus (iii) is proved. $\square$

EXAMPLE 6.3. The procedure described in the preceding theorem for converting A into a reduced row echelon matrix is automatic and can be easily programmed on a computer. However, humans abhor fractions, so we can sometimes make appropriate choices to avoid certain divisions that might occur. Since there is a unique answer, we might as well take a path that is computationally simpler. Consider for example, the integer matrix

$$A_0 = \begin{bmatrix} 0 & 2 & 4 & 1 & 3 \\ 0 & 5 & 10 & 3 & 8 \\ 0 & 3 & 6 & 2 & 2 \end{bmatrix}$$

Dividing by $2, 5$ or 3 will certainly introduce fractions. So we first apply $\mathfrak{R}_3(1, 3; -1)$, namely we subtract the third row from the first. This yields

$$A_1 = \begin{bmatrix} 0 & -1 & -2 & -1 & 1 \\ 0 & 5 & 10 & 3 & 8 \\ 0 & 3 & 6 & 2 & 2 \end{bmatrix}$$

and then multiplying the first row by -1, we have

$$A_2 = \begin{bmatrix} 0 & 1 & 2 & 1 & -1 \\ 0 & 5 & 10 & 3 & 8 \\ 0 & 3 & 6 & 2 & 2 \end{bmatrix}$$

At this point, we subtract 5 times the first row from the second, and then 3 times the first row from the third to obtain

$$A_3 = \begin{bmatrix} 0 & 1 & 2 & 1 & -1 \\ 0 & 0 & 0 & -2 & 13 \\ 0 & 0 & 0 & -1 & 5 \end{bmatrix}$$

In the fourth column, it is certainly better to deal with the -1 entry rather than dividing by 2. So we interchange rows 2 and 3, and then multiply the new second row by -1. This yields

$$A_4 = \begin{bmatrix} 0 & 1 & 2 & 1 & -1 \\ 0 & 0 & 0 & 1 & -5 \\ 0 & 0 & 0 & -2 & 13 \end{bmatrix}$$

Next subtracting the second row from the first, and adding 2 times the second row to the third yields

$$A_5 = \begin{bmatrix} 0 & 1 & 2 & 0 & 4 \\ 0 & 0 & 0 & 1 & -5 \\ 0 & 0 & 0 & 0 & 3 \end{bmatrix}$$

At this point, we have no choice but to divide the last row by 3. Fortunately all the other entries in that row are 0, so again we do not introduce fractions in the matrix

$$A_6 = \begin{bmatrix} 0 & 1 & 2 & 0 & 4 \\ 0 & 0 & 0 & 1 & -5 \\ 0 & 0 & 0 & 0 & 1 \end{bmatrix}$$

Finally, we subtract 4 times the third row from the first, and add 5 times the third row to the second to obtain the reduced row echelon matrix

$$A_7 = \begin{bmatrix} 0 & 1 & 2 & 0 & 0 \\ 0 & 0 & 0 & 1 & 0 \\ 0 & 0 & 0 & 0 & 1 \end{bmatrix}$$

and the procedure is complete.

Problems

6.1. If A is a matrix in row echelon form, prove that the nonzero rows of A are linearly independent. This can be done without serious computation.

6.2. List the analogous elementary column operations and prove the column version of Theorem 6.1.

6.3. Prove that any elementary row operation $\mathfrak{R}$ and any elementary column operation $\mathfrak{C}$ commute in their action on matrices. To be precise, show that $\mathfrak{R}(\mathfrak{C}(A)) = \mathfrak{C}(\mathfrak{R}(A))$ for all matrices A.

6.4. Assume that row subscripts $\{i_1, i_2\}$ and $\{k_1, k_2\}$ are disjoint. Show that the elementary row operations $\mathfrak{R}_3(i_1, k_1; c_1)$ and $\mathfrak{R}_3(i_2, k_2; c_2)$ commute.

6.5. Let $A \in F^{m \times n}$. Using a sequence of elementary row and column operations show that any matrix A can be transformed into a matrix of the form $D_r = [d_{ij}]$, where $d_{11} = d_{22} = \cdots = d_{rr} = 1$ and all the remaining d_{ij} are 0.

6.6. Find an example of an integer matrix whose corresponding reduced row echelon form matrix does not have all integer entries.

6.7. Find a "slanted" basis for the row space of matrix

$$A = \begin{bmatrix} 1 & 1 & 1 & 2 & 0 \\ 2 & 3 & 0 & 1 & 3 \\ 1 & 2 & -1 & -1 & 3 \\ 0 & 1 & -2 & -2 & 1 \end{bmatrix}$$

That is, find a basis that comes from an appropriate reduced row echelon matrix.

6.8. Suppose $A \in F^{n \times n}$ is a square matrix with linearly independent rows. If A is converted into the reduced row echelon form matrix A', find the structure of A'.

6.9. If we "unwrap" matrices into straight horizontal lines, then $F^{m \times n}$ surely looks like the vector space F^{mn}. With this idea in hand, define addition and scalar multiplication so that $F^{m \times n}$ becomes a vector space over F.

6.10. If $F^{m \times n}$ is viewed as a vector space as above, determine its dimension and find a nice basis.

7. Linear Equations and Bases

The elementary row operations are of course familiar because we use them to solve systems of linear equations via *Gaussian elimination*. Thus suppose we are given the system

$$a_{11}x_1 + a_{12}x_2 + \cdots + a_{1n}x_n = b_1$$
$$a_{21}x_1 + a_{22}x_2 + \cdots + a_{2n}x_n = b_2$$
$$\cdots\cdots\cdots$$
$$a_{m1}x_1 + a_{m2}x_2 + \cdots + a_{mn}x_n = b_m$$

of m linear equations in the n unknowns $x_1, x_2, \ldots, x_n$. Of course, we assume that the field elements a_{ij} and b_j are known.

We can clearly group these constants into several appropriate matrices. To start with, we have

$$A = \begin{bmatrix} a_{11} & a_{12} & \cdots & a_{1n} \\ a_{21} & a_{22} & \cdots & a_{2n} \\ & \cdots\cdots\cdots & \\ a_{m1} & a_{m2} & \cdots & a_{mn} \end{bmatrix}, \qquad B = \begin{bmatrix} b_1 \\ b_2 \\ \cdot \\ b_m \end{bmatrix}$$

where A is the *matrix of coefficients* and B is the *matrix of constants*. Notice that the columns of A are labeled $1, 2, \ldots, n$ and these naturally correspond to the n unknowns $x_1, x_2, \ldots, x_n$.

Furthermore, we can combine the two matrices to form

$$A|B = \begin{bmatrix} a_{11} & a_{12} & \cdots & a_{1n} & b_1 \\ a_{21} & a_{22} & \cdots & a_{2n} & b_2 \\ & \cdots\cdots\cdots & & \cdot \\ a_{m1} & a_{m2} & \cdots & a_{mn} & b_m \end{bmatrix}$$

the so-called *augmented matrix* with the constants on the right. The vertical line is of course not formally part of the matrix structure, but it does help us to better visualize how the matrix is partitioned. This matrix clearly contains all the information given by the system of equations, except perhaps for the names of the unknowns. As is to be expected, we have

LEMMA 7.1. *Let $\mathfrak{R}$ be an elementary row operation. Then the set of solutions to the linear system associated to the augmented matrix $A|B$ is the same as the set of solutions to the linear system associated to the augmented matrix $\mathfrak{R}(A|B)$.*

PROOF. Write $\mathfrak{R}(A|B) = A'|B'$ and let $(x_1, x_2, \ldots, x_n) \in F^n$ be a solution to the system of equations

$$a_{i1}x_1 + a_{i2}x_2 + \cdots + a_{in}x_n = b_i$$

associated with $A|B$. In other words, these are honest expressions in F that hold for all $i = 1, 2, \ldots, m$. Obviously these hold if we interchange equations i and k or if we multiply equation i by $c \in F$ to yield

$$(c{\cdot}a_{i1})x_1 + (c{\cdot}a_{i2})x_2 + \cdots + (c{\cdot}a_{in})x_n = c{\cdot}b_i$$

Finally, if we add c times the kth equation to the ith, we get

$$(a_{i1} + c{\cdot}a_{k1})x_1 + (a_{i2} + c{\cdot}a_{k2})x_2 + \cdots + (a_{in} + c{\cdot}a_{kn})x_n = b_i + c{\cdot}b_k$$

It follows that $(x_1, x_2, \ldots, x_n)$ is also a solution of the system associated with $A'|B'$. For the reverse inclusion, we just observe from Lemma 6.1 that $A|B = \mathfrak{R}^{-1}(A'|B')$. $\square$

In view of the above, we are obviously motivated to convert the augmented matrix into reduced row echelon form and to consider the leading 1s.

EXAMPLE 7.1. Suppose the augmented matrix $A|B$ has been converted to $A'|B'$, a matrix in reduced row echelon form and suppose a leading 1, occurs in the last column, the column of constants. Then the corresponding row of $A'|B'$ consists of all 0s followed by a 1 at the end, and the corresponding linear equation is clearly $0 = 1$. Of course, this contradicts (M4) in the definition of a field. It follows that this system of equations is *inconsistent*, that is there are no solutions.

Thus it make sense to assume that the leading 1s occur in the first n columns, namely the columns of A' and of A. But, as we observed, the columns of A correspond to the unknowns $x_1, x_2, \ldots, x_n$ so we can introduce the following notation. We say that the unknowns corresponding to the columns with a leading 1 are the *bound variables* and the remaining unknowns are the *free variables*. The key result here is that solutions always exist and that the free variables can be "freely chosen".

THEOREM 7.1. *Given a system a linear equations with augmented matrix $A|B$. Let $A'|B'$ be the reduced row echelon form matrix obtained from $A|B$, and assume that no leading 1 occurs in the B' column. Then for any choice of free variables in F, there exist unique bound variables in F that yield a solution to the original system of equations.*

PROOF. Since the solution set of the systems corresponding to $A|B$ and to $A'|B'$ are identical, we can assume that $A = A'$ and $B = B'$. Let

$s_1, s_2, \ldots, s_r$ be the column numbers of A corresponding to the leading 1s and let $\mathcal{F}$ denote the set of remaining column numbers. Then since $A|B$ is in reduced row echelon form, the r nonzero equations look like

$$x_{s_i} + \sum_{k \in \mathcal{F}} x_k a_{ik} = b_i$$

for $i = 1, 2, \ldots, r$. In particular, if the free variables are chosen in any way, then the bound variables are uniquely determined by these equations and all such equations are satisfied.   □

In view of Example 3.5, we know that the solution set of a system of homogeneous equations in n unknowns is a subspace of F^n. A close look at the proof of the preceding theorem yields the following more precise formulation of that result.

COROLLARY 7.1. *Given a system of homogeneous linear equations with matrix of coefficients A, and let $A' = [a'_{ij}]$ be the reduced row echelon matrix obtained from A. If $s_1, s_2, \ldots, s_r$ are the column numbers of A' having a leading 1, and if $\mathcal{F}$ is the complementary set of column numbers, then the solution set of the given equations has the basis $\mathcal{A} = \{\alpha_i \mid i \in \mathcal{F}\}$. Here α_i has a 1 as its ith entry, 0 as its kth entry for $i \neq k \in \mathcal{F}$, and $-a'_{ji}$ as its s_jth entry for $j = 1, 2, \ldots, r$.*

PROOF. For each $i \in \mathcal{F}$, let $\alpha_i \in F^n$ be the unique solution of the system of homogenous linear equations associated with A satisfying $x_i = 1$ and $x_k = 0$ for all $i \neq k \in \mathcal{F}$. Then α_i has a 1 as its ith entry and a 0 as its kth entry for all $i \neq k \in \mathcal{F}$. Furthermore, we know that the solution sets for A and for A' are identical, and that the jth nonzero row of A' gives rise to the homogeneous equation

$$x_{s_j} + \sum_{k \in \mathcal{F}} x_k a'_{jk} = 0$$

so $x_{s_j} = -a'_{ji}$. In other words, α_i has $-a'_{ji}$ as its s_jth entry for $j = 1, 2, \ldots, r$ and α_i is now completely understood.

Finally, for all field elements c_i with $i \in \mathcal{F}$, we see that

$$\beta = \sum_{i \in \mathcal{F}} c_i \alpha_i$$

is a solution to the system of homogeneous linear equations with $x_i = c_i$ for all $i \in \mathcal{F}$. Thus, by Theorem 7.1, β is the unique such solution. With this, it follows immediately that $\mathcal{A} = \{\alpha_i \mid i \in \mathcal{F}\}$ is a basis for the solution set.   □

As a consequence, we can easily solve the second problem posed at the beginning of the previous section. Namely, given a finite subset $\mathcal{A} = \{\alpha_1, \alpha_2, \ldots, \alpha_m\}$ of F^n, how do we find a subset of $\mathcal{A}$ that forms a basis for $V = \langle \mathcal{A} \rangle$. At the very least, we should start by checking whether the set $\mathcal{A}$ is linearly independent. Indeed, if this occurs then $\mathcal{A}$ is both a spanning and linearly independent subset of V, and hence a basis for V. Surprisingly, this is all we have to do.

To begin with, let us write

$$\alpha_i = (a_{i1}, a_{i2}, \ldots, a_{in}) \in F^n$$

for $i = 1, 2, \ldots, m$. In the preceding section, we formed the $m \times n$ matrix $A = [a_{ij}]$ and studied its row space. Here we consider the equation

$$x_1 \alpha_1 + x_2 \alpha_2 + \cdots + x_m \alpha_m = 0$$

to see whether $\mathcal{A}$ is linearly independent or not. By checking the n entries in F^n, this gives rise to the system of linear equations

$$a_{11} x_1 + a_{21} x_2 + \cdots + a_{m1} x_m = 0$$
$$a_{12} x_1 + a_{22} x_2 + \cdots + a_{m2} x_m = 0$$
$$\cdots\cdots\cdots$$
$$a_{1n} x_1 + a_{2n} x_2 + \cdots + a_{mn} x_m = 0$$

Now the matrix of coefficients of this system is not the matrix A. Rather its size is $n \times m$, the rows and columns are interchanged and it is the *transpose* of A, namely

$$A^\mathsf{T} = \begin{bmatrix} a_{11} & a_{21} & \cdots & a_{m1} \\ a_{12} & a_{22} & \cdots & a_{m2} \\ & \cdots\cdots\cdots & & \\ a_{1n} & a_{2n} & \cdots & a_{mn} \end{bmatrix}$$

Essentially, if $A = [a_{ij}]$ then $A^\mathsf{T} = [a'_{ij}]$ where $a'_{ij} = a_{ji}$. It is clear that the row space of A^T is identical to the column space of A. Notice that the augmented matrix of the above system of linear equations is $A^\mathsf{T} | Z$ where Z is the column matrix having all entries equal to 0. The following result describes how to find a subset of $\mathcal{A}$ that forms a basis for $V = \langle \mathcal{A} \rangle$.

THEOREM 7.2. *Let $\mathcal{A} = \{\alpha_1, \alpha_2, \ldots, \alpha_m\}$ be a finite subset of F^n and consider the system of n equations in m unknowns determined by*

$$x_1 \alpha_1 + x_2 \alpha_2 + \cdots + x_m \alpha_m = 0$$

Then the vectors α_i corresponding to the bound variables x_i form a basis for $V = \langle \mathcal{A} \rangle \subseteq F^n$. In particular, the dimension of V is equal to the number of bound variables.

PROOF. Suppose the bound variables have subscripts $s_1, s_2, \ldots, s_r$. We have to show that $\mathcal{B} = \{\alpha_{s_1}, \alpha_{s_2}, \ldots, \alpha_{s_r}\}$ is linearly independent and spans $V = \langle \mathcal{A} \rangle$. To start consider

$$x_{s_1}\alpha_{s_1} + x_{s_2}\alpha_{s_2} + \cdots + x_{s_r}\alpha_{s_r} = 0$$

for suitable $x_{s_i} \in F$. Notice that this is a special case of the equation $\sum_j x_j\alpha_j = 0$ but with all free variables equal to 0. Since this equation has the all 0s solution, the uniqueness aspect of Theorem 7.1 implies that all $x_{s_i} = 0$ and hence $\mathcal{B}$ is linearly independent.

To show that $\mathcal{B}$ spans, it clearly suffices to show that each α_k, with x_k a free variable, is contained in the subspace $\langle \mathcal{B} \rangle$. To this end, since we know that we can choose the free variables arbitrarily, let $x_k = -1$ and let all other free variables be 0. Then there exist bound variables x_{s_i} that satisfy the linear system, and the equation $\sum_j x_j\alpha_j = 0$ simplifies to

$$(-1)\alpha_k + \sum_{i=1}^{r} x_{s_i}\alpha_{s_i} = 0$$

But then $\alpha_k = \sum_i x_{s_i}\alpha_{s_i} \in \langle \mathcal{B} \rangle$ and we see that $\mathcal{B}$ spans. It follows that $\mathcal{B}$ is a linearly independent spanning set in V, so $\mathcal{B}$ is indeed a basis for this space. $\qquad\square$

If $A \in F^{m \times n}$, then the row space of A is a subspace of F^n and the column space of A is a subspace of F^m. As a consequence of all of the above, we have the remarkable

COROLLARY 7.2. *If $A \in F^{m \times n}$, then the row space of A and the column space of A have the same dimension. This common dimension is called the rank of A.*

PROOF. Let us start with the matrix A^{T} and via a sequence of elementary row operations, convert it into a reduced row echelon form matrix B. If Z is a column matrix having all 0 entries, then clearly $A^{\mathsf{T}}|Z$ converts to $B|Z$ via the same sequence of operations. Of course, $A^{\mathsf{T}}|Z$ is the augmented matrix for the system of n equations in m unknowns given by

$$x_1\alpha_1 + x_2\alpha_2 + \cdots + x_m\alpha_m = 0$$

where the α_i are the row vectors of A.

Note that $V = \langle \alpha_1, \alpha_2, \ldots, \alpha_m \rangle$ is the row space of A, and by Theorem 7.2, the dimension of V is equal to the number of bound variables x_i and hence equal to the number of leading 1s in B. This is surely equal to the number of nonzero rows of B and hence by Theorem 6.1(ii), this is the dimension of the row space of A^{T}. Since the row space of A^{T} is identical to the column space of A, the result follows. $\qquad\square$

Problems

7.1. Using Gaussian elimination, solve the system of real linear equations given by

$$5x_1 + 5x_2 - x_3 + 7x_4 + 2x_5 + 5x_6 = 9$$
$$x_1 + x_2 - x_3 - x_4 + x_5 - x_6 = 2$$
$$4x_1 + 4x_2 - x_3 + 5x_4 + 2x_5 + 4x_6 = 8$$

Which of the variables are free, which are bound, and what is the solution when all free variables are set to 0?

7.2. Find a basis for the solution space to the system of homogeneous linear equations given by

$$x_1 + 2x_2 + x_3 + 3x_4 + x_5 + x_6 = 0$$
$$2x_1 + 4x_2 + 2x_3 + 6x_4 + 3x_5 + 5x_6 = 0$$
$$x_1 + 2x_2 + x_3 + 2x_4 + 0x_5 + x_6 = 0$$

7.3. Without computation, show that the elementary column operations on A do not change the dimension of the row space of A. Similarly show that the elementary row operations on A do not change the dimension of the column space of A.

7.4. Show that the integer r in Problem 6.5 is uniquely determined. Indeed it is the dimension of the row space of A and of the column space of A.

7.5. Find a subset of the rows of the matrix A of the Problem 6.7 that form a basis for its row space.

7.6. Let A be a square matrix so that A has the same number of rows as columns. Prove that the rows of A are linearly independent if and only if the columns are linearly independent.

7.7. Again let $A \in F^{n \times n}$ be a square matrix and assume that its rows are linearly independent. Consider the system of linear equations associated with $A|B$ and let the reduced row echelon matrix of this augmented matrix be $A'|B'$. Using Problem 6.8, show that B' describes the unique solution to the system of equations.

7.8. State and prove the appropriate analog of Corollary 7.1 in the nonhomogeneous situation.

Let $A \in F^{m \times n}$ and let $B_1, B_2, \ldots, B_t$ be $m \times 1$ column matrices. Consider the system of equations associated with the augmented matrices $A|B_j$ for $j = 1, 2, \ldots, t$ and form the large augmented matrix $A|B$, where B is an $m \times t$ matrix with columns $B_1, B_2, \ldots, B_t$. Via elementary row operations, convert $A|B$ to the reduced row echelon form matrix $A'|B'$.

7.9. Explain how $A'|B'$ can be used to solve the systems associated to the various $A|B_j$ for all j.

7.10. The comments in Example 7.1 do not hold precisely in this context. How can one tell from the matrix $A'|B'$ that the system $A|B_j$ is inconsistent?

CHAPTER II

Linear Transformations

8. Linear Transformations

So far our study of linear algebra has been confined to vector spaces. But in fact the essence of this subject is the study of certain functions defined on these spaces. This is analogous to the situation in calculus. There one first considers the real line, but the main interest of course is in the study of real valued functions defined on the real line.

Let us consider a set of simultaneous linear equations over a field F. Say, for simplicity

$$e_1 x_1 + e_2 x_2 + e_3 x_3 = b$$
$$f_1 x_1 + f_2 x_2 + f_3 x_3 = c$$

We will think of the coefficients $e_1, e_2, e_3, f_1, f_2, f_3$ as being fixed. What we would like to know is for which choices of b and c do solutions exist and then how many solutions are there. Now starting with b and c and finding x_1, x_2, x_3 takes a certain amount of work. On the other hand, starting with x_1, x_2, x_3 and finding b and c from the above is trivial. Of course, if we look at all pairs b, c and all triples x_1, x_2, x_3, then the above two considerations are really the same. Therefore we take the second point of view since it is certainly simpler.

What we have here is a map T that takes ordered triples (x_1, x_2, x_3) with entries in F to ordered pairs (b, c) with entries in F. In other words,

$$T \colon F^3 \to F^2$$

But T is not any old function. It is defined in a linear fashion and therefore we we expect it to have some nice properties.

Let $\alpha = (x_1, x_2, x_3), \beta = (y_1, y_2, y_3) \in F^3$ with $(x_1, x_2, x_3)T = (b, c)$ and $(y_1, y_2, y_3)T = (b', c')$. Observe that we have written this function T to the right of its argument. Then

$$e_1(x_1 + y_1) + e_2(x_2 + y_2) + e_3(x_3 + y_3)$$
$$= (e_1 x_1 + e_2 x_2 + e_3 x_3) + (e_1 y_1 + e_2 y_2 + e_3 y_3)$$
$$= b + b'$$

and similarly

$$f_1(x_1 + y_1) + f_2(x_2 + y_2) + f_3(x_3 + y_3) = c + c'$$

Thus we have

$$(\alpha + \beta)T = (x_1 + y_1, x_2 + y_2, x_3 + y_3)T$$
$$= (b + b', c + c') = \alpha T + \beta T$$

Now let $a \in F$. Then

$$e_1(ax_1) + e_2(ax_2) + e_3(ax_3)$$
$$= a(e_1x_1 + e_2x_2 + e_3x_3) = ab$$

and similarly

$$f_1(ax_1) + f_2(ax_2) + f_3(ax_3) = ac$$

Thus we have

$$(a\alpha)T = (ax_1, ax_2, ax_3)T$$
$$= (ab, ac) = a(\alpha T)$$

and therefore T is a linear transformation.

Let V and W be two vector spaces over the same field F. A *linear transformation* from V to W is a function $T\colon V \to W$ satisfying

T1. For all $\alpha, \beta \in V$, we have

$$(\alpha + \beta)T = \alpha T + \beta T$$

T2. For all $\alpha \in V$ and $a \in F$, we have

$$(a\alpha)T = a(\alpha T)$$

In other words, T somehow intertwines the arithmetic of V and of W. We have already seen that the function T as defined earlier is indeed a linear transformation. We now consider some more examples.

EXAMPLE 8.1. For any two vector spaces V, W over F there is always the map $T\colon V \to W$ given by $\alpha T = 0$ for all $\alpha \in V$. This is clearly a linear transformation which we denote, as expected, by 0.

EXAMPLE 8.2. The identity map $I\colon V \to V$ is a linear transformation. Here of course $\alpha I = \alpha$ for all $\alpha \in V$.

EXAMPLE 8.3. Let W be a subspace of V. Then there is a natural map, the injection of W into V, given by $\alpha T = \alpha$ for all $\alpha \in W$. Observe that the second α is considered to be a vector in V. T is of course a linear transformation since W and V share the same arithmetic.

In the other direction, suppose that W has a complement W' in V so that $V = W \oplus W'$. Then each element $\alpha \in V$ can be written uniquely as $\beta + \beta'$ with $\beta \in W$ and $\beta' \in W'$. This means that the projection map $P\colon V \to W$ given by $\alpha P = \beta$ is well defined, and then it is easily seen to be a linear transformation. We remark that different complements for W give rise to different projection maps.

EXAMPLE 8.4. Let V be a vector space over the field F. For each $a \in F$ define $T_a \colon V \to V$ by $\alpha T_a = a\alpha$ for all $\alpha \in V$. It follows from the distributive and associative laws that T_a is a linear transformation. Observe that $T_0 = 0$ and that, by the unital law, $T_1 = I$.

EXAMPLE 8.5. Let $\mathbb{C}$ be considered as a vector space over $\mathbb{R}$. For each complex number α, let $\mathfrak{Re}(\alpha)$ denote its real part and $\mathfrak{Im}(\alpha)$ its imaginary part. Then both $\mathfrak{Re} \colon \mathbb{C} \to \mathbb{R}$ and $\mathfrak{Im} \colon \mathbb{C} \to \mathbb{R}$ are linear transformations of real vector spaces.

EXAMPLE 8.6. Let C denote the vector space of real valued continuous functions on the interval $[0, 1]$. Let $S \colon C \to \mathbb{R}$ be defined by

$$f(x)S = \int_0^1 f(t)\,dt$$

for all $f(x) \in C$. Then S is a linear transformation from C to $\mathbb{R}$.

EXAMPLE 8.7. Let C be as above and define $J \colon C \to C$ by

$$(fJ)(x) = \int_0^x f(t)\,dt$$

Then J is easily seen to be a linear transformation.

EXAMPLE 8.8. Given integers $m, n \geq 1$, we define $T \colon F^n \to F^m$ as follows. First fix mn elements $a_{ij} \in F$ with $i = 1, 2, \ldots, m$ and $j = 1, 2, \ldots, n$. Then set

$$(c_1, c_2, \ldots, c_n)T = \left(\sum_{j=1}^n a_{1j}c_j, \sum_{j=1}^n a_{2j}c_j, \ldots, \sum_{j=1}^n a_{mj}c_j \right)$$

This is of course a generalization of our original example and it is surely a linear transformation.

EXAMPLE 8.9. Let V be the vector space of all functions from a set U into the field F. Fix $u_0 \in U$. Then the evaluation map $E_0 \colon V \to F$ given by $\alpha E_0 = \alpha(u_0)$ is a linear transformation. More generally, suppose $u_1, u_2, \ldots, u_n \in U$. Then we can define the map $E \colon V \to F^n$ by

$$\alpha E = (\alpha(u_1), \alpha(u_2), \ldots, \alpha(u_n))$$

and E is a linear transformation.

EXAMPLE 8.10. Finally let $V = F[x]$. We can define the formal derivative $D \colon V \to V$ by

$$\left(\sum_{i=0}^\infty a_i x^i \right) D = \sum_{i=1}^\infty i a_i x^{i-1}$$

It is easy to see that D is a linear transformation which enjoys most of the properties of ordinary differentiation. In particular, one can show that

$$(\alpha\beta)D = (\alpha D)\beta + \alpha(\beta D)$$

Before we consider a general means of constructing such maps, let us first note a few elementary properties of linear transformations.

LEMMA 8.1. *Let $T\colon V \to W$ be a linear transformation. Then for all $a_i \in F$ and $\alpha_i \in V$ we have*

$$(a_1\alpha_1 + a_2\alpha_2 + \cdots + a_k\alpha_k)T$$
$$= a_1(\alpha_1 T) + a_2(\alpha_2 T) + \cdots + a_k(\alpha_k T)$$

Moreover $0T = 0$ and $(-\alpha)T = -(\alpha T)$ for all $\alpha \in V$.

PROOF. It follows easily by induction that for $\beta_i \in V$ we have

$$(\beta_1 + \beta_2 + \cdots + \beta_k)T = \beta_1 T + \beta_2 T + \cdots + \beta_k T$$

Thus

$$(a_1\alpha_1 + a_2\alpha_2 + \cdots + a_k\alpha_k)T = (a_1\alpha_1)T + (a_2\alpha_2)T + \cdots + (a_k\alpha_k)T$$
$$= a_1(\alpha_1 T) + a_2(\alpha_2 T) + \cdots + a_k(\alpha_k T)$$

Now let $\alpha \in V$. Then $0 = 0\alpha$ so

$$0T = (0\alpha)T = 0(\alpha T) = 0$$

Finally

$$(-\alpha)T = ((-1)\alpha)T = (-1)(\alpha T) = -(\alpha T)$$

and the lemma is proved. $\square$

We show now that linear transformations are plentiful.

THEOREM 8.1. *Let V be a finite dimensional vector space with basis $\mathcal{B} = \{\beta_1, \beta_2, \ldots, \beta_n\}$ and let W be another space over the same field F. Then for any choice of $\gamma_1, \gamma_2, \ldots, \gamma_n \in W$, there exists one and only one linear transformation $T\colon V \to W$ with $\beta_i T = \gamma_i$ for $i = 1, 2, \ldots, n$. In fact, T is given by*

$$(a_1\beta_1 + a_2\beta_2 + \cdots + a_n\beta_n)T = a_1\gamma_1 + a_2\gamma_2 + \cdots + a_n\gamma_n$$

for all $a_1, a_2, \ldots, a_n \in F$.

PROOF. Suppose $T\colon V \to W$ is given with $\beta_i T = \gamma_i$ for all i. Then, by the previous lemma,

$$(a_1\beta_1 + a_2\beta_2 + \cdots + a_n\beta_n)T$$
$$= a_1(\beta_1 T) + a_2(\beta_2 T) + \cdots + a_n(\beta_n T)$$
$$= a_1\gamma_1 + a_2\gamma_2 + \cdots + a_n\gamma_n$$

Now $\mathcal{B}$ spans V, so every element $\alpha \in V$ can be written as an F-linear sum of the β_i's. Thus the above formula gives αT for all α, and therefore T is uniquely determined.

Now given $\gamma_1, \gamma_2, \ldots, \gamma_n \in W$, we show that T exists. Obviously, there is only one way to define T in view of the above and therefore we set

$$\alpha T = a_1\gamma_1 + a_2\gamma_2 + \cdots + a_n\gamma_n$$

for $\alpha = a_1\beta_1 + a_2\beta_2 + \cdots + a_n\beta_n \in V$. Note that $\mathcal{B}$ is not only a spanning set but it is in fact a basis for V. That means that $\alpha \in V$ can be written in only one way as above and therefore T is a well defined function from V to W. It is now a simple matter to see that T is the required linear transformation.

First, suppose

$$\alpha = a_1\beta_1 + a_2\beta_2 + \cdots + a_n\beta_n$$

and

$$\alpha' = a_1'\beta_1 + a_2'\beta_2 + \cdots + a_n'\beta_n$$

are vectors in V and let $b \in F$. Then

$$\alpha + \alpha' = (a_1 + a_1')\beta_1 + (a_2 + a_2')\beta_2 + \cdots + (a_n + a_n')\beta_n$$

and

$$b\alpha = (ba_1)\beta_1 + (ba_2)\beta_2 + \cdots + (ba_n)\beta_n$$

so by definition

$$(\alpha + \alpha')T = (a_1 + a_1')\gamma_1 + (a_2 + a_2')\gamma_2 + \cdots + (a_n + a_n')\gamma_n$$
$$= (a_1\gamma_1 + a_2\gamma_2 + \cdots + a_n\gamma_n) + (a_1'\gamma_1 + a_2'\gamma_2 + \cdots + a_n'\gamma_n)$$
$$= \alpha T + \alpha' T$$

and

$$(b\alpha)T = (ba_1)\gamma_1 + (ba_2)\gamma_2 + \cdots + (ba_n)\gamma_n$$
$$= b(a_1\gamma_1 + a_2\gamma_2 + \cdots + a_n\gamma_n) = b(\alpha T)$$

Thus T is a linear transformation. Finally,

$$\beta_i = 0\beta_1 + 0\beta_2 + \cdots + 1\beta_i + \cdots + 0\beta_n$$

so
$$\beta_i T = 0\gamma_1 + 0\gamma_2 + \cdots + 1\gamma_i + \cdots + 0\gamma_n = \gamma_i$$
and the theorem is proved. $\square$

There are a number of ways of constructing new linear transformations from old ones. We consider one such now.

Let $T: V \to W$ be a linear transformation. As usual we say that T is *one-to-one* if for each $\gamma \in W$ there exists at most one $\alpha \in V$ with $\alpha T = \gamma$. We say that T is *onto* if for each $\gamma \in W$ there exists at least one $\alpha \in V$ with $\alpha T = \gamma$. Thus if T is one-to-one and onto, then for each $\gamma \in W$ there exists one and only one $\alpha \in V$ with $\alpha T = \gamma$. But this means that α is really a function of γ and we can therefore define naturally a back map $T^{-1}: W \to V$ given by
$$\gamma T^{-1} = \alpha \qquad \text{for} \quad \gamma \in W$$
where α is the unique element of V with $\alpha T = \gamma$. As we might expect, T^{-1} is also a linear transformation.

For brevity we call such a one-to-one and onto linear transformation an *isomorphism*.

LEMMA 8.2. *Let $T: V \to W$ be an isomorphism. Then the map $T^{-1}: W \to V$ is also an isomorphism.*

PROOF. Let $\gamma_1, \gamma_2 \in W$ and let $a \in F$ with $\gamma_1 T^{-1} = \alpha_1$ and $\gamma_2 T^{-1} = \alpha_2$. Then by definition $\alpha_1 T = \gamma_1$ and $\alpha_2 T = \gamma_2$. Since T is a linear transformation, this yields
$$(\alpha_1 + \alpha_2)T = \alpha_1 T + \alpha_2 T = \gamma_1 + \gamma_2$$
$$(a\alpha_1)T = a(\alpha_1 T) = a\gamma_1$$
Thus again by definition of T^{-1} we have
$$(\gamma_1 + \gamma_2)T^{-1} = \alpha_1 + \alpha_2 = \gamma_1 T^{-1} + \gamma_2 T^{-1}$$
$$(a\gamma_1)T^{-1} = a\alpha_1 = a(\gamma_1 T^{-1})$$
and T^{-1} is a linear transformation.

Suppose now that $\gamma_1 T^{-1} = \gamma_2 T^{-1}$. Then $\alpha_1 = \alpha_2$ so
$$\gamma_1 = \alpha_1 T = \alpha_2 T = \gamma_2$$
and hence T^{-1} is one-to-one. Finally, let $\alpha \in V$ and set $\gamma = \alpha T$. Then by definition of T^{-1} we have $\gamma T^{-1} = \alpha$. Thus T^{-1} is onto and the result follows. $\square$

Problems

8.1. Verify that the projection map P of Example 8.3 is a linear transformation.

8.2. Verify that the map T_a given in Example 8.4 is a linear transformation.

8.3. If you already know calculus, convince yourself that the maps S and J given in Examples 8.6 and 8.7 are linear transformations.

8.4. Let $D : F[x] \to F[x]$ be the formal derivative map. Show that

$$(\alpha\beta)D = \alpha(\beta D) + (\alpha D)\beta$$

for all $\alpha, \beta \in F[x]$. (Hint. First verify this for $\alpha = ax^n$ and $\beta = bx^m$, and then prove the result by induction on $\deg \alpha + \deg \beta$.)

8.5. Describe geometrically the linear transformations $T_i \colon \mathbb{R}^2 \to \mathbb{R}^2$ given by

$$(a, b)T_1 = (2a, 2b)$$
$$(a, b)T_2 = (2a, b)$$
$$(a, b)T_3 = (-a, -b)$$

For each T_i find a nonzero vector α_i and a scalar $c_i \in \mathbb{R}$ with $\alpha_i T_i = c_i \alpha_i$.

8.6. Describe geometrically the linear transformation $S_\theta \colon \mathbb{R}^2 \to \mathbb{R}^2$ given by

$$(a, b)S_\theta = (a\cos\theta - b\sin\theta, a\sin\theta + b\cos\theta)$$

For which angles θ can one find $0 \neq \alpha \in \mathbb{R}^2$ and $a \in \mathbb{R}$ with $\alpha S_\theta = a\alpha$.

8.7. Let $T \colon F^3 \to F^2$ be given by

$$(a_1, a_2, a_3)T = (a_1 - a_2, a_3 - 2a_2 + a_1)$$

Show that T is onto, but not one-to-one.

8.8. Let $T \colon F^3 \to F^4$ be given by

$$(a_1, a_2, a_3)T = (a_1 - 2a_2 + a_3, a_1 + a_2, a_2 - a_3, a_3)$$

Prove that T is one-to-one, but not onto.

8.9. Let V be an n-dimensional vector space over F with basis $\mathcal{B} = \{\beta_1, \beta_2, \ldots, \beta_n\}$. Show that the map $T \colon F^n \to V$ given by

$$(a_1, a_2, \ldots, a_n)T = a_1\beta_1 + a_2\beta_2 + \cdots + a_n\beta_n$$

is an isomorphism. Find T^{-1}.

8.10. Let V be a finite dimensional vector space and let W_1 and W_2 be subspaces of the same dimension. Construct an isomorphism $T \colon W_1 \to W_2$ such that $\gamma T = \gamma$ for all $\gamma \in W_1 \cap W_2$.

9. Kernels and Images

Let $T\colon V \to W$ be a linear transformation so that T is a map that intertwines the arithmetic of V and of W. It is apparent therefore that T must preserve certain aspects of the vector space structure. For example, we have already seen that T maps 0 to 0 and negatives to negatives. It is then natural to ask how it behaves with respect to subspaces, bases and dimension.

If V_0 is a subset of V, we let $(V_0)T$ denote the image of V_0, that is

$$(V_0)T = \{\alpha T \mid \alpha \in V_0\}$$

In the same way, if W_0 is a subset of W, we let

$$(W_0)\overleftarrow{T} = \{\alpha \in V \mid \alpha T \in W_0\}$$

denote the complete inverse image of the subset W_0. Thus $(V_0)T \subseteq W$ and $(W_0)\overleftarrow{T} \subseteq V$.

THEOREM 9.1. *Let $T\colon V \to W$ be a linear transformation with V and W vector spaces over the same field F.*

> *i. If V' is a subspace of V, then $(V')T$ is a subspace of W.*
> *ii. If W' is a subspace of W, then $(W')\overleftarrow{T}$ is a subspace of V.*

PROOF. Recall that in order to verify that a subset is a subspace, it suffices to check that the set is closed under addition and scalar multiplication and that it contains 0.

(i) Let $\beta_1, \beta_2 \in (V')T$ and let $a \in F$. By definition, there exist $\alpha_1, \alpha_2 \in V'$ with $\beta_1 = \alpha_1 T$ and $\beta_2 = \alpha_2 T$. Then $\alpha_1 + \alpha_2 \in V'$ and $a\alpha_1 \in V'$ so

$$\beta_1 + \beta_2 = (\alpha_1 + \alpha_2)T \in (V')T$$
$$a\beta_1 = (a\alpha_1)T \in (V')T$$

Finally $0 = 0T \in (V')T$, so $(V')T$ is a subspace of W.

(ii) Let $\alpha_1, \alpha_2 \in (W')\overleftarrow{T}$ and let $a \in F$. Then by definition, $\alpha_1 T, \alpha_2 T \in W'$ so since W' is a subspace of W, we have

$$(\alpha_1 + \alpha_2)T = \alpha_1 T + \alpha_2 T \in W'$$
$$(a\alpha_1)T = a(\alpha_1 T) \in W'$$

Hence, by definition, $\alpha_1 + \alpha_2, a\alpha_1 \in (W')\overleftarrow{T}$. Finally, $0T = 0 \in W'$, so $0 \in (W')\overleftarrow{T}$ and $(W')\overleftarrow{T}$ is a subspace of V. $\qquad\square$

Now a vector space always has two subspaces of interest namely itself and 0. This gives rise to the following definitions.

Let $T\colon V \to W$ be a linear transformation. We set

$$kernel \text{ of } T = \ker T = (0)\overleftarrow{T}$$
$$image \text{ of } T = \operatorname{im} T = (V)T$$

Then $\ker T$ is a subspace of V and $\operatorname{im} T$ is a subspace of W.

THEOREM 9.2. *Let $T\colon V \to W$ be a linear transformation.*

    i. Let $\alpha_1, \alpha_2 \in V$. Then $\alpha_1 T = \alpha_2 T$ if and only if $\alpha_1 - \alpha_2 \in \ker T$.

    ii. T is one-to-one if and only if $\ker T = 0$.

    iii. T is onto if and only if $\operatorname{im} T = W$.

PROOF. (i) Since T is a linear transformation, we have $\alpha_1 T - \alpha_2 T = (\alpha_1 - \alpha_2)T$. Thus $\alpha_1 T = \alpha_2 T$ if and only if $(\alpha_1 - \alpha_2)T = 0$ and hence if and only if $\alpha_1 - \alpha_2 \in \ker T$.

(ii) Suppose T is one-to-one. Then $\ker T = (0)\overleftarrow{T}$ must consist of at most one vector. Since $0 \in \ker T$ this yields $\ker T = 0$. Conversely suppose $\ker T = 0$. If $\alpha_1 T = \alpha_2 T$, then $\alpha_1 - \alpha_2 \in \ker T = 0$, so $\alpha_1 = \alpha_2$.

(iii) This follows by definition. $\qquad\qquad\qquad\qquad\qquad\qquad\Box$

Let us consider some examples.

EXAMPLE 9.1. We start with some trivialities. The zero map $0\colon V \to W$ certainly satisfies $\ker 0 = V$ and $\operatorname{im} 0 = 0$. Also the identity map $I\colon V \to V$ satisfies $\ker I = 0$, $\operatorname{im} I = V$.

EXAMPLE 9.2. Suppose that V is the real vector space $V = \mathbb{R}[x]$. If $D\colon V \to V$ is the derivative map given by

$$\left(\sum_{i=0}^{\infty} a_i x^i\right) D = \sum_{i=1}^{\infty} i a_i x^{i-1}$$

then clearly $\ker D$ is the set of constant polynomials and $\operatorname{im} D = \mathbb{R}[x]$. Thus D is onto but not one-to-one. Moreover we conclude, as is well known, that two polynomials have the same derivative if and only if they differ by an element of $\ker D$, namely a constant.

Now let $S\colon V \to V$ denote the integral map given by

$$\left(\sum_{i=0}^{\infty} a_i x^i\right) S = \sum_{i=0}^{\infty} \frac{a_i}{i+1} x^{i+1}$$

Then $\ker S = 0$ and $\operatorname{im} S$ is the set of all polynomials with constant term 0. In particular, S is one-to-one but not onto.

EXAMPLE 9.3. Let V be a finite dimensional vector space and let W be a subspace. Then there exists a complementary subspace W' for W so that $V = W \oplus W'$. Let $P'\colon V \to W'$ be the projection map given by $\alpha P' = \beta'$ where α is uniquely written as $\alpha = \beta + \beta'$ with $\beta \in W$ and $\beta' \in W'$. Clearly $\operatorname{im} P' = W$ and $\ker P' = W$. The latter fact shows that any subspace of V can be the kernel of some linear transformation. Thus being the kernel of a linear transformation is no more special than being a subspace.

We now consider bases and dimension. The main result of this section is

THEOREM 9.3. *Let* $T \colon V \to W$ *be a linear transformation with* V *and* W *both finite dimensional over* F. *Then there exist bases* $\mathcal{B} = \{\alpha_1, \ldots, \alpha_r, \beta_1, \ldots, \beta_s\}$ *of* V *and* $\mathcal{C} = \{\gamma_1, \ldots, \gamma_s, \delta_1, \ldots, \delta_t\}$ *of* W *such that*

> i. $\{\alpha_1, \ldots, \alpha_r\}$ *is a basis for* $\ker T$.
> ii. $\beta_j T = \gamma_j$ *for* $j = 1, \ldots, s$.
> iii. $\{\gamma_1, \ldots, \gamma_s\}$ *is a basis for* $\operatorname{im} T$.

Thus

$$\dim_F \ker T + \dim_F \operatorname{im} T = \dim_F V$$

PROOF. Since $\ker T$ is a subspace of V, it has a basis $\{\alpha_1, \ldots, \alpha_r\}$ that extends to a basis $\mathcal{B} = \{\alpha_1, \ldots, \alpha_r, \beta_1, \ldots, \beta_s\}$ of V. Set $\gamma_1 = \beta_1 T, \gamma_2 = \beta_2 T, \ldots, \gamma_s = \beta_s T$. We show that $\{\gamma_1, \ldots, \gamma_s\}$ is a basis for $\operatorname{im} T$.

First let $\gamma \in \operatorname{im} T$. Then $\gamma = \alpha T$ for some $\alpha \in V$ and since $\mathcal{B}$ is a basis for V we have

$$\alpha = a_1 \alpha_1 + \cdots + a_r \alpha_r + b_1 \beta_1 + \cdots + b_s \beta_s$$

for suitable $a_i, b_i \in F$. Thus

$$\gamma = \alpha T = a_1(\alpha_1 T) + \cdots + a_r(\alpha_r T) + b_1(\beta_1 T) + \cdots + b_s(\beta_s T)$$
$$= b_1 \gamma_1 + \cdots + b_s \gamma_s$$

since $\alpha_i \in \ker T$ implies that $\alpha_i T = 0$. We have therefore shown that $\{\gamma_1, \ldots, \gamma_s\}$ spans $\operatorname{im} T$.

Now suppose $b_1 \gamma_1 + \cdots + b_s \gamma_s = 0$ for some $b_i \in F$. Then

$$(b_1 \beta_1 + \cdots + b_s \beta_s)T = b_1(\beta_1 T) + \cdots + b_s(\beta_s T)$$
$$= b_1 \gamma_1 + \cdots + b_s \gamma_s = 0$$

so $b_1 \beta_1 + \cdots + b_s \beta_s \in \ker T$. Now $\{\alpha_1, \ldots, \alpha_r\}$ spans $\ker T$ so

$$b_1 \gamma_1 + \cdots + b_s \gamma_s = a_1 \alpha_1 + \cdots + a_r \alpha_r$$

for suitable $a_i \in F$. This yields

$$(-a_1)\alpha_1 + \cdots + (-a_r)\alpha_r + b_1 \gamma_1 + \cdots + b_s \gamma_s = 0$$

so since $\mathcal{B}$ is a basis for V all coefficients must be zero. In particular, $b_1 = \cdots = b_s = 0$ so $\{\gamma_1, \ldots, \gamma_s\}$ is linearly independent and therefore a basis for $\operatorname{im} T$. We now extend $\{\gamma_1, \ldots, \gamma_s\}$ to $\mathcal{C} = \{\gamma_1, \ldots, \gamma_s, \delta_1, \ldots, \delta_t\}$, a basis for W.

Finally, $\dim_F \ker T = r$ and $\dim_F \operatorname{im} T = s$, so

$$\dim_F V = r + s = \dim_F \ker T + \dim_F \operatorname{im} T$$

and the result follows. □

If $T\colon V \to W$, then we define the *rank* of T to be the dimension of the image of T. Therefore a restatement of the last part of the previous theorem is

COROLLARY 9.1. *Let* $T\colon V \to W$ *be a linear transformation. Then*

$$\dim_F \ker T + \operatorname{rank} T = \dim_F V$$

Recall that a linear transformation $T\colon V \to W$ is said to be an isomorphism if it is one-to-one and onto.

COROLLARY 9.2. *If* $T\colon V \to W$ *is an isomorphism of finite dimensional vector spaces, then* $\dim_F V = \dim_F W$.

PROOF. This is clear from Theorem 9.3 since $\dim_F \ker T = 0$ and $\dim_F \operatorname{im} T = \dim_F W$. □

Next, we consider a linear transformation T from V to V. Such a map T is said to be *nonsingular* if it is an isomorphism.

COROLLARY 9.3. *Let* V *be a vector space of dimension* $n < \infty$ *and let* $T\colon V \to V$ *be a linear transformation. Then the following are equivalent.*

    i. T is nonsingular.
    ii. T is onto, that is $\operatorname{rank} T = n$.
    iii. T is one-to-one.

PROOF. Clearly (i) implies (ii) and (iii). In the other direction, we use

$$\dim_F \ker T + \operatorname{rank} T = \dim_F V$$

If T is onto, then $\operatorname{rank} T = \dim_F V = n$, so $\dim_F \ker T = 0$. This shows that $\ker T = 0$, so T is also one-to-one and hence nonsingular. Finally, if T is one-to-one, then $\ker T = 0$, so $\operatorname{rank} T = n$. Since $\dim_F \operatorname{im} T = \operatorname{rank} T = \dim_F V$, we have $\operatorname{im} T = V$ so T is onto and the result follows. □

We observed in Example 9.2 that with $V = \mathbb{R}[x]$, the linear transformation D is onto but not one-to-one. Also S is one-to-one but not onto. Thus the above corollary does not hold in general without the finite dimensionality assumption.

COROLLARY 9.4. *Let* $T\colon V \to W$ *be a linear transformation with* $\dim_F W < \dim_F V < \infty$. *Then* $\ker T \neq 0$.

PROOF. We have

$$\dim_F V = \dim_F \operatorname{im} T + \dim_F \ker T$$

Now $\operatorname{im} T$ is a subspace of W, so $\dim_F \operatorname{im} T \leq \dim_F W < \dim_F V$. This easily yields $\dim_F \ker T > 0$, so $\ker T \neq 0$. $\qquad\square$

EXAMPLE 9.4. Let us see now how all this applies to the study of linear equations. We consider a set of m linear equations in the n unknowns $x_1, x_2, \ldots, x_n$. This is given by

$$a_{11}x_1 + a_{12}x_2 + \cdots + a_{1n}x_n = b_1$$
$$a_{21}x_1 + a_{22}x_2 + \cdots + a_{2n}x_n = b_2$$
$$\cdots\cdots$$
$$a_{m1}x_1 + a_{m2}x_2 + \cdots + a_{mn}x_n = b_m$$

Observe that, as usual, the coefficients $a_{ij} \in F$ are double subscripted. The first subscript corresponds to the row or the equation, and the second subscript corresponds to the column of the unknown.

We think of the set $\{a_{ij}\}$ as being fixed. Then the above equations define a linear transformation $T\colon F^n \to F^m$ given by

$$(x_1, x_2, \ldots, x_n)T = (b_1, b_2, \ldots, b_m)$$

Now the solution set of the *homogeneous* equations, that is when $b_1 = b_2 = \cdots = b_m = 0$ is clearly $(0)T$, the kernel of T. This is of course a subspace of F^n and hence has a basis, say $\{\alpha_1, \alpha_2, \ldots, \alpha_r\}$. Thus every solution of the system of homogeneous equations can be written uniquely as

$$a_1\alpha_1 + a_2\alpha_2 + \cdots + a_r\alpha_r$$

for suitable $a_i \in F$. See Corollary 7.2 for efficient ways to construct such bases.

Now we ask for which $(b_1, b_2, \ldots, b_m) \in F^m$ do solutions exist. But clearly a solution exists if and only if $(b_1, b_2, \ldots, b_m) \in \operatorname{im} T$. In particular, the set of m-tuples of constant terms for which a solution exists is in fact a subspace of F^m. Suppose $(b_1, b_2, \ldots, b_m) \in \operatorname{im} T$ and let $\beta = (y_1, y_2, \ldots, y_n)$ be a solution to the associated equations so that $\beta T = (b_1, b_2, \ldots, b_m)$. Then β' is also a solution if and only if $\beta' - \beta \in \ker T$ or in other words if and only if

$$\beta' = \beta + a_1\alpha_1 + a_2\alpha_2 + \cdots + a_r\alpha_r$$

for suitable $a_i \in F$. Thus once we know one solution for a particular vector in $\operatorname{im} T$, we can find all such. In addition the above formula shows that, in some sense, all such vectors in $\operatorname{im} T$ give rise to the same "number" of solutions.

Finally let us consider two applications of the preceding theorems. First suppose $n > m$. Then

$$\dim_F F^n = n > m = \dim_F F^m$$

and hence $\ker T \neq 0$. This says that if there are more unknowns than equations, then there is always a nonzero solution to the set of homogeneous equations.

Secondly, suppose $m = n$. Then T is one-to-one if and only if it is onto. This says that a solution always exists if and only if solutions that do exist are always unique.

Problems

Let $T \colon V \to W$ be a linear transformation with V and W both vector spaces over F.

9.1. Suppose T is onto. Prove that T and $\overleftarrow{T}$ define a one-to-one correspondence between subspaces of V that contain $\ker T$ and all subspaces of W.

9.2. Let V_1 be a subspace of V and let W_1 be a subspace of W with $(V_1)T \subseteq W_1$. Show that the restriction map $T_1 \colon V_1 \to W_1$ given by $\alpha T_1 = \alpha T$ for $\alpha \in V_1$ is a linear transformation. Moreover show that $\ker T_1 = V_1 \cap (\ker T)$.

9.3. Suppose T is onto and let V_1 be a complement for $\ker T$ in V. Prove that the restriction map $T_1 \colon V_1 \to W$ is an isomorphism.

Consider the following set of three linear equations in four unknowns with coefficients in $\mathbb{Q}$.

$$2x_1 + x_2 + x_3 - 4x_4 = b_1$$
$$3x_1 + x_2 + 3x_3 - x_4 = b_2$$
$$x_1 + x_2 - x_3 - 7x_4 = b_3$$

and let $T \colon \mathbb{Q}^4 \to \mathbb{Q}^3$ be the corresponding linear transformation.

9.4. Find a basis for $\ker T$ and extend this to a basis for $\mathbb{Q}^4$.

9.5. Find a basis for $\operatorname{im} T$ and extend this to a basis for $\mathbb{Q}^3$. What is the rank of T?

9.6. Find all solutions with $b_1 = 2$, $b_2 = 3$, $b_3 = 1$.

9.7. Let V be an F-vector space of dimension $n < \infty$ and let $T: V \to V$ be a linear transformation. Define subspaces V_j and W_j inductively by

$$V_0 = V, \qquad V_{j+1} = (V_j)T$$
$$W_0 = 0, \qquad W_{j+1} = (W_j)\overleftarrow{T}$$

Show that $V_j \supseteq V_{j+1}$ and $W_{j+1} \supseteq W_j$ and deduce from this that $V_{2n} = V_n$ and $W_{2n} = W_n$.

Let V be a two-dimensional real vector space with basis $\{\alpha_1, \alpha_2\}$. Let $T: V \to V$ be the linear transformation given by

$$\alpha_1 T = 4\alpha_1 - 5\alpha_2$$
$$\alpha_2 T = 2\alpha_1 - 3\alpha_2$$

9.8. Find nonzero vectors $\beta_1, \beta_2 \in V$ with $\beta_1 T = -\beta_1$ and $\beta_2 T = 2\beta_2$.

9.9. Suppose $0 \neq \gamma \in V$ with $\gamma T = a\gamma$ for some $a \in \mathbb{R}$. Show that $a = -1$ or 2.

9.10. Prove that $\{\beta_1.\beta_2\}$ is a basis for V and describe T as above in terms of this basis.

10. Quotient Spaces

In Example 9.3, we showed that if V is a finite dimensional vector space, then every subspace of V is the kernel of some linear transformation $T \colon V \to V'$. This fact is also true for infinite dimensional vector spaces and can be proved using the transfinite methods we will study at the end of these notes. But there is an alternate approach that is quite elementary and avoids these infinite methods. Indeed, it applies to many other algebraic systems where complements do not exist in general. We consider this below.

To start with, we need a better understanding of equality. For example, what does it mean when we say

$$\frac{2}{3} = \frac{4}{6}$$

Surely the fractions are not identical since they have different numerators and different denominators. So equality here must mean "equal in $\mathbb{Q}$" or perhaps "numerically equal". Again $4 - 3 = 1$ must mean "equal in the integers $\mathbb{Z}$" or perhaps "numerically equal". Even worse, what does $3 = 3$ mean? Surely the two 3's are not absolutely identical and certainly they are composed of different molecules. Again $=$ must mean "numerically equal".

Let $\mathcal{A}$ be a set and let $\sim$ be a relation on $\mathcal{A}$. In other words, for any two elements of $a, b \in \mathcal{A}$, we have either $a \sim b$ or not. We say that $\sim$ is an *equivalence relation* if

$\quad \mathcal{E}1.\ a \sim a$ for all $a \in A$. $\hfill$ (reflexive)
$\quad \mathcal{E}2.\ a \sim b$ if and only if $b \sim a$. $\hfill$ (symmetric)
$\quad \mathcal{E}3.$ If $a \sim b$ and $b \sim c$, then $a \sim c$. $\hfill$ (transitive)

Our first example is amusing but surprisingly relevant.

EXAMPLE 10.1. Suppose we have a large collection of marbles and we store them in a number of different buckets. Let us say that marbles a and b are equivalent if and only if they are stored in the same bucket. Certainly $a \sim a$, and $a \sim b$ if and only if $b \sim a$. Finally, since b is in a unique bucket, we see that $a \sim b$ and $b \sim c$ imply that a and c are in the same bucket as b and hence $a \sim c$.

EXAMPLE 10.2. We can of course formulate the above example without the use of marbles. Let $\mathcal{A}$ be a set and suppose $\mathcal{A}$ is given as

$$\mathcal{A} = \dot{\bigcup}_{i \in \mathcal{I}} \mathcal{A}_i$$

a disjoint union of the nonempty subsets $\mathcal{A}_i$. If $a, b \in \mathcal{A}$, we say that a and b are equivalent, and write $a \sim b$, if and only if a and b are in the same subset $\mathcal{A}_i$. As above, it is easy to verify that $\sim$ is an equivalence relation.

Let us return to Example 10.1 and ask, which bucket contains marble a? We cannot answer this exactly, since we don't really know what a bucket is, but we can say that the bucket we are looking for is the one that contains all the marbles that are equivalent to a. So if we identify the bucket with its contents, then we can say that the bucket is equal to $\{x \mid x \sim a\}$. This motivates the following definition.

Let $\mathcal{A}$ be a set with an equivalence relation $\sim$. For each $a \in \mathcal{A}$ we define the *equivalence class* of a to be

$$\mathrm{cl}(a) = \{x \in \mathcal{A} \mid x \sim a\}$$

and we write $\mathcal{A}/\sim$ for the set of all these equivalence classes. In particular, we have a natural map $\mathrm{cl} \colon \mathcal{A} \to \mathcal{A}/\sim$ that is certainly onto. With this, we obtain a converse to Example 10.2.

THEOREM 10.1. *If $\mathcal{A}$ is a set with an equivalence relation $\sim$, then $\mathcal{A}$ is the disjoint union of its distinct equivalence classes.*

PROOF. Notice that $a \in \mathrm{cl}(a)$ so $\mathcal{A}$ is certainly equal to the union of its equivalence classes. We need only show that distinct classes are disjoint. To start with, we observe that if $a \sim b$, then $\mathrm{cl}(a) = \mathrm{cl}(b)$. Indeed, if $x \in \mathrm{cl}(a)$, then by transitivity $x \sim a$ and $a \sim b$ imply that $x \sim b$ and hence $x \in \mathrm{cl}(b)$. Thus $\mathrm{cl}(a) \subseteq \mathrm{cl}(b)$ and by symmetry we have the reverse inclusion. Finally, if $\mathrm{cl}(a) \cap \mathrm{cl}(b) \neq \emptyset$ and if c is in the intersection, then $c \sim a$ and $c \sim b$. It follows from symmetry and transitivity that $a \sim b$ and hence $\mathrm{cl}(a) = \mathrm{cl}(b)$, as required. □

Before we apply this to our vector space problem, it is certainly worthwhile to consider other examples of interest. The first is well known, but uses a different symbol for equivalence.

EXAMPLE 10.3. Let $\mathcal{A} = \mathbb{Z}$ be the ring of rational integers and let $n > 1$ be a fixed integer. Then $n\mathbb{Z}$ is the set of all $\mathbb{Z}$-multiples of n, namely those integers divisible by n. If $a, b \in \mathbb{Z}$, we write $a \equiv b \bmod n$ if and only if n divides $a - b$. Then $\equiv$ is an equivalence relation. Indeed, $a - a = 0 = n \cdot 0$ so $a \equiv a \bmod n$, and if $a - b$ is divisible by n, then so is $(b - a) = -(a - b)$. Finally, if $a \equiv b \bmod n$ and if $b \equiv c \bmod n$, then $a - b$ is a multiple of n and $b - c$ is a multiple of n. Hence $a - c = (a - b) + (b - c)$ is a multiple of n and $a \equiv c \bmod n$.

Now if $a \in \mathbb{Z}$ then $\mathrm{cl}(a)$ is easily seen to be the *coset* $a + n\mathbb{Z}$ and the distinct classes are $\mathrm{cl}(0) = n\mathbb{Z}$, $\mathrm{cl}(1) = 1 + n\mathbb{Z}$, ..., through $\mathrm{cl}(n-1) = (n-1) + n\mathbb{Z}$. Indeed, for $i = 0, 1, \ldots, n-1$, we see that $\mathrm{cl}(i) = i + n\mathbb{Z}$ is precisely the set of all integers that leave a remainder of i when divided by n.

Next, we look at the fraction problem we began with.

EXAMPLE 10.4. Let us assume that we understand the integers $\mathbb{Z}$ and we want to understand $\mathbb{Q}$. To start with, we consider all "formal fractions" a/b with $a, b \in \mathbb{Z}$ and $b \neq 0$. Observe that these are just ordered pairs (a, b), where we know the numerator a and the denominator b. Denote by $\mathcal{F}$ the set of all such fractions and write $a/b \sim c/d$ if and only if $ad = bc$, the formula one gets by cross multiplying. Notice that the latter is an equation in $\mathbb{Z}$ and hence well understood. We show now that $\sim$ is an equivalence relation. Indeed, since multiplication in $\mathbb{Z}$ is commutative, it follows immediately that $\sim$ is reflexive and symmetric. For transitivity, suppose $a/b \sim c/d$ and $c/d \sim e/f$. Then $ad = bc$ and $cf = de$, so

$$d(af - be) = daf - dbe$$
$$= f(ad - bc) + b(cf - de) = 0$$

Thus, since $d \neq 0$, we have $af - be = 0$ and hence $a/b \sim e/f$.

So, what is $\mathbb{Q}$? Well, if $q \in \mathbb{Q}$, then $q = a/b$ for some $a, b \in \mathbb{Z}$ with $b \neq 0$. Furthermore, q is also equal to another fraction c/d if and only if the formal fractions a/b and c/d are equivalent. Thus q really corresponds to the equivalence class of a/b. In other words, there is a one-to-one correspondence, that is a one-to-one, onto map, $\theta \colon \mathbb{Q} \to \mathcal{F}/\sim$. Indeed, when $\mathbb{Q}$ is constructed from $\mathbb{Z}$, it is taken to be $\mathcal{F}/\sim$. Of course, $\mathbb{Q} \supseteq \mathbb{Z}$ and $\mathbb{Q}$ has an arithmetic defined on it. So these aspects require additional considerations. We will outline some of this in Problems 10.5 through 10.8.

The moral here is that we can define a new structure by starting with something we know, introducing an appropriate equivalence relation, and then studying the set of equivalence classes. With this idea in hand, we move on to consider vector spaces. As will be apparent, the situation here is quite similar to that of Example 10.3 except that we expand upon it by introducing an arithmetic on the set of classes.

Let V be a vector space over the field F and let W be a fixed subspace. For $\alpha, \beta \in V$ we write $\alpha \sim \beta$, or more properly $\alpha \sim_W \beta$ if and only if $\alpha - \beta \in W$.

LEMMA 10.1. *With the above notation, we have*

  i. $\sim_W$ *is an equivalence relation on* V.
  ii. If $\alpha \in V$, *then* $\mathrm{cl}(\alpha)$ *is equal to the coset* $\alpha + W$.
  iii. If $\alpha \sim \alpha'$ *and* $\beta \sim \beta'$, *then* $\alpha + \beta \sim \alpha' + \beta'$.
  iv. If $\alpha \sim \alpha'$ *and* $c \in F$, *then* $c\alpha \sim c\alpha'$.

Parts (iii) and (iv) above say that the equivalence relation *respects the arithmetic* in V. As we will see, results of this nature allow us to define an appropriate arithmetic on the set of equivalence classes.

PROOF. (i) It is interesting to observe that the three conditions for $\sim$ to be an equivalence relation almost precisely mirror the three conditions in Theorem 3.1 for W to be a subspace. First, if $\alpha \in V$, then $\alpha - \alpha = 0 \in W$, so $\alpha \sim \alpha$. Next suppose $\alpha \sim \beta$. Then $\alpha - \beta \in W$, so $\beta - \alpha = (-1)(\alpha - \beta) \in W$ and $\beta \sim \alpha$. Finally, if $\alpha \sim \beta$ and $\beta \sim \gamma$, then $\alpha - \beta \in W$ and $\beta - \gamma \in W$, so $\alpha - \gamma = (\alpha - \beta) + (\beta - \gamma) \in W$ and $\alpha \sim \gamma$.

(ii) For this, we note that $\beta \sim \alpha$ if and only $\beta - \alpha \in W$ and hence if and only if $\beta \in \alpha + W$.

(iii) If $\alpha \sim \alpha'$ and $\beta \sim \beta'$, then $\alpha - \alpha' \in W$ and $\beta - \beta' \in W$ so $(\alpha + \beta) - (\alpha' + \beta') = (\alpha - \alpha') + (\beta - \beta') \in W$ and $\alpha + \beta \sim \alpha' + \beta'$.

(iv) Finally, if $\alpha \sim \alpha'$ and $c \in F$, then $\alpha - \alpha' \in W$ so $c\alpha - c\alpha' = c(\alpha - \alpha') \in W$ and $c\alpha \sim c\alpha'$. $\square$

Now let us return to Example 10.1 to better visualize what comes next. Let us assume that the collection of marbles in that example actually has an arithmetic, say a multiplication $*$ defined on it. Can we extend this in a natural manner to a multiplication of the buckets? The obvious approach is as follows. Let $\mathcal{A}_1$ and $\mathcal{A}_2$ be buckets and choose a marble $a_1 \in \mathcal{A}_1$ and $a_2 \in \mathcal{A}_2$. Then we can multiply a_1 and a_2, see which bucket $\mathcal{B}$ contains $a_1 * a_2$, and define $\mathcal{A}_1 * \mathcal{A}_2$ to equal $\mathcal{B}$. This seems reasonable, but there is an obvious flaw. Namely, suppose we choose a second marble a_1' in $\mathcal{A}_1$ and a second marble a_2' in $\mathcal{A}_2$. We then multiply these new elements and see which bucket $\mathcal{B}'$ contains $a_1' * a_2'$. If $\mathcal{B}'$ is always equal to $\mathcal{B}$, then everything is fine and the definition makes sense. If not, the process just does not work.

So how can we guarantee that all $\mathcal{B}'$ equal $\mathcal{B}$? Note that a_1 and a_1' come from the same bucket, so $a_1 \sim a_1'$. Similarly $a_2 \sim a_2'$. Obviously we need some sort of result which asserts that for all choices of marbles, $a_1 \sim a_1'$ and $a_2 \sim a_2'$ implies that $(a_1 * a_2) \sim (a_1' * a_2')$. Indeed, if we have such a result, then $a_1 * a_2$ and $a_1' * a_2'$ will always belong to the same bucket. In other words, we need to show that the relation $\sim$ respects the multiplication $*$. When this occurs we can indeed define the multiplication of buckets, and since $\mathcal{A}_1 = \mathrm{cl}(a_1)$ and $\mathcal{A}_2 = \mathrm{cl}(a_2)$, we obtain $\mathrm{cl}(a_1) * \mathrm{cl}(a_2) = \mathrm{cl}(a_1 * a_2)$.

Now as we observed, parts (iii) and (iv) of the preceding lemma show that in a vector space V, the relation $\sim_W$ does indeed respect its arithmetic. Thus, we can define addition and scalar multiplication in the set of equivalence classes of V, and then show that $V/\sim$ is also an F-vector space. Once we do this, we write $V/\sim$ as V/W, the *quotient space* of V by W.

THEOREM 10.2. *Let W be a subspace of the F-vector space V and let $\sim$ be the equivalence relation on V determined by W. Then*

    i. $V/W = V/\sim$ is an F-vector space.

    ii. The class map $T = \mathrm{cl}$ is a linear transformation from V onto the quotient space V/W.

    iii. $\ker T = W$.

In particular, every subspace W of V is the kernel of a suitable linear transformation.

PROOF. Since the equivalence relation $\sim_W$ respects addition and scalar multiplication by the preceding lemma, we can define a corresponding addition and scalar multiplication in $V/\sim$. Indeed, if we write T for the class map $T : V \to V/\sim$, then $\alpha T + \beta T = (\alpha + \beta)T$ and $c(\alpha T) = (c\alpha)T$ for all $\alpha, \beta \in V$ and $c \in F$. In particular, if we knew that $V/\sim$ was an F-vector space, that is if it satisfied all the appropriate axioms, then T would be a linear transformation.

Now we know at least that T is onto, and with this we can use T to transfer the axioms from V to $V/\sim$. To start with, we already know that closure of addition and of scalar multiplication are satisfied. Next come the *for all* axioms. These are the identities that hold for all elements of V and F. For example, addition is associative in V so $\alpha + (\beta + \gamma) = (\alpha + \beta) + \gamma$ for all $\alpha, \beta, \gamma \in V$. Thus, applying T to this expression yields

$$\alpha T + (\beta T + \gamma T) = \alpha T + (\beta + \gamma)T = (\alpha + (\beta + \gamma))T$$
$$= ((\alpha + \beta) + \gamma)T = (\alpha + \beta)T + \gamma T$$
$$= (\alpha T + \beta T) + \gamma T$$

But T is onto, so $\alpha T, \beta T$ and γT are three typical elements of $V/\sim$, and thus we conclude that the associative law of addition holds in $V/\sim$.

Similarly, the distributive law $(c + d)\alpha = c\alpha + d\alpha$ holds in V, and applying T yields

$$(c + d)(\alpha T) = ((c + d)\alpha)T = (c\alpha + d\alpha)T$$
$$= (c\alpha)T + (d\alpha)T = c(\alpha T) + d(\alpha T)$$

Again, T is onto, so αT is a typical element of $V/\sim$ and therefore this distributive law holds in $V/\sim$. The remaining axioms of this nature obviously carry over in the same way.

Thus, we need only consider the zero axiom and negatives. Since $\alpha + 0 = 0 + \alpha = \alpha$, we see that

$$\alpha T + 0T = 0T + \alpha T = \alpha T$$

Thus since αT is a typical element of $V/\sim$, we see that $0T$ plays the role of 0 in $V/\sim$. In the same way, $(-\alpha)T$ plays the role of $-(\alpha T)$ since $\alpha + (-\alpha) = (-\alpha) + \alpha = 0$ yields

$$\alpha T + (-\alpha)T = (-\alpha)T + \alpha T = 0T = 0$$

Thus $V/\sim$ is indeed a vector space over F and $T\colon V \to V/\sim$ is an onto linear transformation. Finally, since $0 = 0T$, we see that $\alpha \in \ker T$ if and only if $\mathrm{cl}(\alpha) = \alpha T = 0 = 0T = \mathrm{cl}(0)$. Obviously, this occurs if and only if $\alpha \sim 0$ and hence if and only if $\alpha - 0 \in W$. Thus $\ker T = W$ and the theorem is proved. $\qquad\square$

Let us return, one last time, to Example 10.1 and suppose that we have a function f that assigns to each marble some attribute, say its color. Can we use f to assign a color to each bucket? Obviously, we can do this if all marbles in the same bucket have the same color. In other words, we need the equivalence relation $\sim$ to *respect the function* f, so that $a \sim b$ implies $af = bf$. Since the bucket containing a is the equivalence class of a, we can then define a function $\bar{f}$ on buckets that sends $\mathrm{cl}(a)$ to af.

Finally, we apply this idea to our vector space situation.

THEOREM 10.3. *Let W be a subspace of the F-vector space V and let $S\colon V \to V'$ be a linear transformation with $\ker S \supseteq W$. Then there is a natural linear transformation $\bar{S}\colon V/W \to V'$ such that*

  i. For all $\alpha \in V$,

$$\alpha S = (\alpha T)\bar{S}$$

  where $T\colon V \to V/W$ is the class map.

  ii. $\operatorname{im} \bar{S} = \operatorname{im} S$ and hence $\bar{S}$ is onto if and only if S is onto.

  iii. $\alpha T \in \ker \bar{S}$ if and only if $\alpha \in \ker S$. Hence $\bar{S}$ is one-to-one if and only if $\ker S = W$.

PROOF. Let $\sim$ denote the equivalence relation on V determined by W, and let $T\colon V \to V/W$ be the class map. We show first that $\sim$ respects the function S, and this is where the hypothesis $\ker S \supseteq W$ is used. Indeed, if $\alpha \sim \beta$, then $\alpha - \beta \in W \subseteq \ker S$, so $\alpha S - \beta S = (\alpha - \beta)S = 0$ and $\alpha S = \beta S$. As above this allows us to define a map $\bar{S}\colon V/W \to V'$ by

$$(\alpha T)\bar{S} = (\operatorname{cl}(\alpha))\bar{S} = \alpha S$$

We can now quickly verify that $\bar{S}$ is a linear transformation. To this end, let $\alpha, \beta \in V$ and $c \in F$. Then, since both S and T are linear transformations, we have

$$(\alpha T + \beta T)\bar{S} = ((\alpha + \beta)T)\bar{S} = (\alpha + \beta)S$$
$$= \alpha S + \beta S = (\alpha T)\bar{S} + (\beta T)\bar{S}$$

and

$$(c(\alpha T))\bar{S} = ((c\alpha)T)\bar{S} = (c\alpha)S$$
$$= c(\alpha S) = c((\alpha T)\bar{S})$$

As usual, we note that T is onto so αT and βT are two typical elements of V/W. With this, we conclude that $\bar{S}$ is indeed a linear transformation.

Finally, from $(\alpha T)\bar{S} = \alpha S$ we see that $\operatorname{im} \bar{S} = \operatorname{im} S$. Thus $\bar{S}$ is onto if and only if S is onto. Furthermore, $\alpha T \in \ker \bar{S}$ if and only if $\alpha S = 0$ and hence if and only if $\alpha \in \ker S$. In particular, $\bar{S}$ is one-to-one if and only if $(\ker S)T = 0$ and therefore if and only if $\ker S \subseteq \ker T = W$. $\quad\square$

Recall that an isomorphism of vector spaces is a linear transformation that is a one-to-one correspondence. Namely, it is one-to-one and onto.

COROLLARY 10.1. *Let V be an F-vector space and let $S\colon V \to V'$ be a linear transformation onto V'. If $W = \ker S$, then $\bar{S}\colon V/W \to V'$ is an isomorphism of vector spaces.*

PROOF. We know that $\bar{S}$ is a linear transformation, and parts (ii) and (iii) of the preceding theorem imply that $\bar{S}$ is onto and one-to-one. Thus $\bar{S}$ is an isomorphism. $\quad\square$

Problems

10.1. In Example 10.2, where is the disjointness of the various subsets $\mathcal{A}_i$ used in the proof that $\sim$ is an equivalence relation. In particular, show that transitivity fails if the subsets are not disjoint.

10.2. Example 10.3 studies $\mathcal{A} = \mathbb{Z}$ and congruence modulo $n > 1$. Show that $\equiv$ respects the arithmetic of $\mathbb{Z}$. In other words, verify that if $a \equiv a'$ and $b \equiv b'$ then $a + b \equiv a' + b'$ and $ab \equiv a'b'$. Sketch a proof that $\mathbb{Z}/n\mathbb{Z} = \mathbb{Z}/{\equiv}$ is a *commutative ring*, namely it satisfies all the field axioms except the one concerning the existence of multiplicative inverses.

10.3. If n above is not a prime number show that the ring $\mathbb{Z}/n\mathbb{Z}$ contains two nonzero elements that multiply to 0. Prove that neither of these elements can have an inverse, and conclude that the ring is not a field.

10.4. Notice that $\mathbb{Z}/2\mathbb{Z}$ is the finite field with two elements that is described in Example 1.4. Now it is known that if $a, b \in \mathbb{Z}$ are relatively prime, that is they have no nontrivial factors in common, then there exist $x, y \in \mathbb{Z}$ with $ax + by = 1$. Use this to prove that for any prime p, the ring $\mathbb{Z}/p\mathbb{Z}$ is a finite field of size p.

10.5. Let $\mathcal{F}$ be the set of all formal integer fractions as described in Example 10.4 and let $\sim$ be the corresponding equivalence relation. Define an arithmetic on $\mathcal{F}$ by

$$(a/b) + (c/d) = (ad + bc)/(bd)$$
$$(a/b) \cdot (c/d) = (ac)/(bd)$$

Show that $\sim$ respects this arithmetic.

10.6. Prove that the addition and multiplication defined above are commutative and associative. Show that $\mathbb{Z}$ embeds in $\mathcal{F}$ via the map $\varphi \colon a \mapsto a/1$ and that φ preserves addition and multiplication.

10.7. Show that the distributive law fails in $\mathcal{F}$. Prove instead that

$$(a/b) \cdot \big((c/d) + (e/f) \big) \sim (a/b) \cdot (c/d) + (a/b) \cdot (e/f)$$

10.8. Starting with the known ring $\mathbb{Z}$, use these ideas to construct the rational field $\mathbb{Q}$. Sketch a proof that $\mathbb{Q}$ satisfies all the appropriate field axioms, that $\mathbb{Q}$ contains a copy $\mathbb{Z}$ and that every element of $\mathbb{Q}$ is a fraction with numerator and denominator in this copy of $\mathbb{Z}$.

10.9. Let $V = W \oplus W'$ be a vector space over F and let $P' \colon V \to W'$ denote the corresponding projection map described in Example 9.3. If $\sim$ is the equivalence relation on V determined by W, show that the linear transformation $\overline{P'} \colon V/W \to W'$, as given by Theorem 10.3, is an isomorphism that assigns to each equivalence class of V the unique element of W' it contains.

10.10. Let $U \subseteq W \subseteq V$ be a chain of vector spaces and write $S \colon V \to V/U$ and $T \colon V \to V/W$ for the corresponding class maps. Let $S_0 \colon W \to V/U$ denote the restriction of S to W, as described in Problem 9.2, and observe, by Theorem 10.3, that $\overline{S_0} \colon W/U \to V/U$ is an isomorphism to its image $(W)S$. Now show that T gives rise to a linear transformation $\overline{T} \colon V/U \to V/W$ that is onto and has kernel equal to the image of $\overline{S_0}$.

11. Matrix Correspondence

We have already made a good deal of progress on the solving aspects of linear equations. We observed for example that the solution set of a system of homogeneous equations has defined on it a natural arithmetic, and this led to the definition of a vector space. We also showed that these solution sets had finite bases and therefore that all elements in it could be easily described. There is of course still more to do, but for now we look in another direction. We consider the systems of linear equations or rather their associated linear transformations and we describe a natural arithmetic on them.

Let V and W be two vector spaces over the same field F and let $\mathcal{L}(V, W)$ denote the set of all linear transformations from V to W. Suppose $S, T \in \mathcal{L}(V, W)$ and $a \in F$. We can then define the maps $S + T$ and aS by

$$\alpha(S + T) = \alpha S + \alpha T$$
$$\alpha(aS) = a(\alpha S) \qquad \text{for all } \alpha \in V$$

We observe now that these maps are also linear transformations from V to W.

Let $\alpha, \beta \in V$ and $b \in F$. Then by definition and the fact that both S and T are linear transformations, we have

$$\begin{aligned}
(\alpha + \beta)(S + T) &= (\alpha + \beta)S + (\alpha + \beta)T \\
&= (\alpha S + \beta S) + (\alpha T + \beta T) \\
&= (\alpha S + \alpha T) + (\beta S + \beta T) \\
&= \alpha(S + T) + \beta(S + T)
\end{aligned}$$

$$\begin{aligned}
(b\alpha)(S + T) &= (b\alpha)S + (b\alpha)T \\
&= b(\alpha S) + b(\alpha T) \\
&= b(\alpha S + \alpha T) = b \cdot \alpha(S + T)
\end{aligned}$$

and similarly

$$\begin{aligned}
(\alpha + \beta)(aS) &= a \cdot (\alpha + \beta)S \\
&= a(\alpha S + \beta S) \\
&= a(\alpha S) + a(\beta S) \\
&= \alpha(aS) + \beta(aS)
\end{aligned}$$

$$(b\alpha)(aS) = a\cdot(b\alpha)S$$
$$= a\cdot b(\alpha S) = b\cdot a(\alpha S)$$
$$= b\cdot\alpha(aS)$$

Thus $S + T \in \mathcal{L}(V, W)$ and $aS \in \mathcal{L}(V, W)$. In other words, $\mathcal{L}(V, W)$ is a set with an addition and a scalar multiplication defined on it. It is in fact a vector space over F.

THEOREM 11.1. *Let V and W be vector spaces over F. Then with addition and scalar multiplication defined as above, $\mathcal{L}(V, W)$ is also a vector space over F.*

PROOF. We have already verified the closure axioms. The associative, commutative, distributive and unitary laws are routine to check and so we relegate these to Problem 11.1 at the end of this section. Finally the zero map clearly plays the role of $0 \in \mathcal{L}(V, W)$ and $-T$ is just $(-1)T$ as defined above. The result follows.   □

At this point we could go ahead and compute $\dim_F \mathcal{L}(V, W)$ by constructing a basis. However we take another approach.

Let m and n be positive integers and consider the space F^{mn}. This is of course the vector space of mn-tuples over F. Now it doesn't really matter how we write these mn entries, namely whether we write them in a straight line or in a circle or perhaps in a rectangular m by n array. So for reasons that will be apparent later on we will take this latter approach and write all such elements as m by n arrays. When we do this, we of course designate F^{mn} by $F^{m\times n}$ and we call this the space of $m \times n$ (m by n) *matrices* over F. We have seen these matrices before in Sections 6 and 7 as formal arrays, but without considering their arithmetic.

Now, if $\alpha \in F^{m\times n}$, then α is an m by n array of elements of F, where m indicates the number of rows and n the number of columns. As usual, we write α as

$$\alpha = \begin{bmatrix} a_{11} & a_{12} & \cdots & a_{1n} \\ a_{21} & a_{22} & \cdots & a_{2n} \\ & \cdots\cdots\cdots & & \\ a_{m1} & a_{m2} & \cdots & a_{mn} \end{bmatrix}$$

in terms of double subscripted entries $a_{ij} \in F$. Here i is the row subscript and runs between 1 and m, and j is the column subscript and runs between 1 and n. For brevity we sometimes denote the *matrix α* by $\alpha = \begin{bmatrix} a_{ij} \end{bmatrix}$.

Since $F^{m\times n}$ is really just F^{mn} in disguise, we know how addition and scalar multiplication are defined. Never-the-less to avoid confusion we restate this below. Let $\alpha = [a_{ij}]$ and $\beta = [b_{ij}]$ be matrices in $F^{m \times n}$ and let $c \in F$. Then

$$\alpha + \beta = [a_{ij}] + [b_{ij}] = [a_{ij} + b_{ij}]$$

and

$$c\alpha = c[a_{ij}] = [ca_{ij}]$$

In other words, the i, j-th entry of $\alpha + \beta$ is $a_{ij} + b_{ij}$ and the i, j-th entry of $c\alpha$ is ca_{ij}. Clearly

$$\dim_F F^{m \times n} = \dim_F F^{mn} = mn$$

Let us return now to $\mathcal{L}(V, W)$. Suppose $\dim_F V = m < \infty$, $\dim_F W = n < \infty$ and choose bases $\mathcal{A} = \{\alpha_1, \alpha_2, \ldots, \alpha_m\}$ of V and $\mathcal{B} = \{\beta_1, \beta_2, \ldots, \beta_n\}$ of W. Here we think of $\mathcal{A}$ and $\mathcal{B}$ as not only sets, but in fact ordered sets, namely the basis vectors are suitably numbered $1, 2, \ldots$. Let $T \in \mathcal{L}(V, W)$. Then $\alpha_i T \in W$, so $\alpha_i T$ can be written uniquely as an F-linear sum of the β's. This then gives rise to the family of equations

$$\alpha_1 T = a_{11}\beta_1 + a_{12}\beta_2 + \cdots + a_{1n}\beta_n$$
$$\alpha_2 T = a_{21}\beta_1 + a_{22}\beta_2 + \cdots + a_{2n}\beta_n$$
$$\cdots\cdots$$
$$\alpha_m T = a_{m1}\beta_1 + a_{m2}\beta_2 + \cdots + a_{mn}\beta_n$$

for uniquely determined elements $a_{ij} \in F$. We then have a map which we denote by $_{\mathcal{A}}[\ ]_{\mathcal{B}}$ from $\mathcal{L}(V, W)$ to $F^{m \times n}$ given by

$$_{\mathcal{A}}[\ ]_{\mathcal{B}} \colon T \mapsto \begin{bmatrix} a_{11} & a_{12} & \cdots & a_{1n} \\ a_{21} & a_{22} & \cdots & a_{2n} \\ \multicolumn{4}{c}{\cdots\cdots\cdots} \\ a_{m1} & a_{m2} & \cdots & a_{mn} \end{bmatrix}$$

where the a_{ij} are as above. Unlike our usual function on the right notation, we will use $_{\mathcal{A}}[T]_{\mathcal{B}}$ to denote the image of T under this map. Now $\mathcal{L}(V, W)$ and $F^{m \times n}$ are both vector spaces over F, and we have

THEOREM 11.2. *Let V and W be finite dimensional vector spaces over the field F with $\mathcal{A} = \{\alpha_1, \alpha_2, \ldots, \alpha_m\}$ being a basis for V and $\mathcal{B} = \{\beta_1, \beta_2, \ldots, \beta_n\}$ being a basis for W. Then*

$$_{\mathcal{A}}[\ ]_{\mathcal{B}} \colon \mathcal{L}(V, W) \to F^{m \times n}$$

as defined above is a rank preserving isomorphism. In particular,

$$\dim_F \mathcal{L}(V, W) = mn = (\dim_F V)(\dim_F W)$$

and $\operatorname{rank} T = \operatorname{rank} {}_{\mathcal{A}}[T]_{\mathcal{B}}$ for all $T \in \mathcal{L}(V, W)$.

PROOF. We first show that $_\mathcal{A}[\ ]_\mathcal{B}$ is a linear transformation. Let $T, T' \in \mathcal{L}(V, W)$ with $_\mathcal{A}[T]_\mathcal{B} = [a_{ij}]$ and $_\mathcal{A}[T']_\mathcal{B} = [a'_{ij}]$ and let $b \in F$. Then by definition of $T + T'$ and bT we have for all i

$$\alpha_i(T + T') = \alpha_i T + \alpha_i T'$$
$$= (a_{i1}\beta_1 + a_{i2}\beta_2 + \cdots + a_{in}\beta_n)$$
$$+ (a'_{i1}\beta_1 + a'_{i2}\beta_2 + \cdots + a'_{in}\beta_n)$$
$$= (a_{i1} + a'_{i1})\beta_1 + (a_{i2} + a'_{i2})\beta_2 + \cdots + (a_{in} + a'_{in})\beta_n$$

and similarly

$$\alpha_i(bT) = b(\alpha_i T)$$
$$= b(a_{i1}\beta_1 + a_{i2}\beta_2 + \cdots + a_{in}\beta_n)$$
$$= (ba_{i1})\beta_1 + (ba_{i2})\beta_2 + \cdots + (ba_{in})\beta_n$$

This implies that

$$_\mathcal{A}[T + T']_\mathcal{B} = [a_{ij} + a'_{ij}]$$
$$= [a_{ij}] + [a'_{ij}] = {}_\mathcal{A}[T]_\mathcal{B} + {}_\mathcal{A}[T']_\mathcal{B}$$

and

$$_\mathcal{A}[bT]_\mathcal{B} = [ba_{ij}]$$
$$= b[a_{ij}] = b \cdot {}_\mathcal{A}[T]_\mathcal{B}$$

Therefore $_\mathcal{A}[\ ]_\mathcal{B}$ is a linear transformation.

Recall that we have shown in Theorem 8.1 that for each choice of $\gamma_1, \gamma_2, \ldots, \gamma_m \in W$ there exists one and only one linear transformation $T \colon V \to W$ with $\alpha_i T = \gamma_i$ for $i = 1, 2, \ldots, m$. Since the γ_i are uniquely determined by

$$\gamma_i = a_{i1}\beta_1 + a_{i2}\beta_2 + \cdots + a_{in}\beta_n$$

we see immediately that $_\mathcal{A}[\ ]_\mathcal{B}$ is one-to-one and onto. Thus $_\mathcal{A}[\ ]_\mathcal{B}$ is an isomorphism, and by Corollary 9.2 we conclude that $\dim_F \mathcal{L}(V, W) = \dim_F F^{m \times n} = mn$.

It remains to compare the ranks of T and of $A = {}_\mathcal{A}[T]_\mathcal{B} = [a_{ij}]$ and we sketch a proof of their equality. First observe that the map $S \colon F^n \to W$ given by

$$(b_1, b_2, \ldots, b_n) \mapsto b_1\beta_1 + b_2\beta_2 + \cdots + b_n\beta_n$$

is an isomorphism since $\mathcal{B} = \{\beta_1, \beta_2, \ldots, \beta_n\}$ is a basis for W. Next note from Corollary 7.2 that the rank of A is the dimension of its row space, namely the subspace of F^n spanned by the row vectors

$\rho_i = (a_{i1}, a_{i2}, \ldots, a_{in})$ for $i = 1, 2, \ldots, m$. Furthermore, the rank of T is the dimension of $\operatorname{im} T$, the subspace of W spanned by the vectors

$$\alpha_i T = a_{i1}\beta_1 + a_{i2}\beta_2 + \cdots + a_{in}\beta_n = \rho_i S$$

again for $i = 1, 2, \ldots, m$. Since S is an isomorphism, it induces an isomorphism of the subspaces

$$S \colon \langle \rho_1, \rho_2, \ldots, \rho_m \rangle \to \langle \rho_1 S, \rho_2 S, \ldots, \rho_m S \rangle$$
$$= \langle \alpha_1 T, \alpha_2 T, \ldots, \alpha_m T \rangle = \operatorname{im} T$$

and Corollary 9.2 implies that

$$\operatorname{rank} A = \dim_F \langle \rho_1, \rho_2, \ldots, \rho_m \rangle = \dim_F \operatorname{im} T = \operatorname{rank} T$$

as required.   □

We call $_{\mathcal{A}}[T]_{\mathcal{B}}$ the matrix associated with T. Obviously different choices of bases $\mathcal{A}$ and $\mathcal{B}$ give rise to different matrices.

We close this section with a slightly different phenomenon. Namely we study a multiplication that can be defined between certain pairs of matrices. Let us consider two matrices $\alpha = [a_{ij}]$ and $\beta = [b_{ij}]$ over F but of possibly different sizes. Say α is $m \times n$ and β is $r \times s$. That is symbolically

$$\alpha = m \overset{n}{\begin{bmatrix} \\ \\ \end{bmatrix}}, \qquad \beta = r \overset{s}{\begin{bmatrix} \\ \\ \end{bmatrix}}$$

Then the product $\alpha\beta$ is defined if and only if $n = r$ in which case $\alpha\beta$ turns out to be of size $m \times s$. In other words,

$$m \overset{\not{n}}{\begin{bmatrix} \\ \\ \end{bmatrix}} \cdot \not{n} \overset{s}{\begin{bmatrix} \\ \\ \end{bmatrix}} = m \overset{s}{\begin{bmatrix} \\ \\ \end{bmatrix}}$$

and we can think of the intermediate n's as somehow being cancelled.

Now $\alpha\beta = [c_{ij}]$ has the same number of rows as does α and the same number of columns as does β. Thus it would make sense to have c_{ij} be a function of the ith row of α and the jth column of β. In addition, the rows of α and the columns of β have the same number of entries, namely n. Therefore we can and do define

$$c_{ij} = \sum_{k=1}^{n} a_{ik} b_{kj}$$

We see that this meets all of the above criteria and therefore at least makes sense.

We repeat the definition formally. Let $\alpha = \left[a_{ij}\right]$ be an $m \times n$ matrix over F and let $\beta = \left[b_{ij}\right]$ be an $r \times s$ matrix over F. Then the product $\alpha\beta = \left[c_{ij}\right]$ is defined if and only if $n = r$ in which case $\alpha\beta$ is $m \times s$ and

$$c_{ij} = \sum_{k=1}^{n=r} a_{ik}b_{kj}$$

for all appropriate i and j.

It is natural to ask whether this multiplication has reasonably nice properties. The answer is "yes and no". First observe that if $\alpha\beta$ is defined, then $\beta\alpha$ need not be, so we cannot expect a general commutative law. However even when both $\alpha\beta$ and $\beta\alpha$ are defined and have the same size, we still need not have $\alpha\beta = \beta\alpha$. For example, if

$$\alpha = \begin{bmatrix} 1 & 0 \\ 0 & 0 \end{bmatrix} \quad \text{and} \quad \beta = \begin{bmatrix} 0 & 1 \\ 0 & 0 \end{bmatrix}$$

then

$$\alpha\beta = \begin{bmatrix} 0 & 1 \\ 0 & 0 \end{bmatrix} \quad \text{and} \quad \beta\alpha = \begin{bmatrix} 0 & 0 \\ 0 & 0 \end{bmatrix}$$

so $\alpha\beta \neq \beta\alpha$ in this case. Now the sum of two matrices is defined if and only if they have the same size. With this understanding, we have the following

LEMMA 11.1. *Matrix multiplication is associative and distributive. That is, for all matrices α, β, γ over F of suitable sizes (so that the arithmetic operations are defined) we have*

$$\alpha(\beta\gamma) = (\alpha\beta)\gamma$$
$$\alpha(\beta + \gamma) = \alpha\beta + \alpha\gamma$$
$$(\beta + \gamma)\alpha = \beta\alpha + \gamma\alpha$$

PROOF. We first consider the associative law. Suppose that the left hand side is defined. Then $\beta\gamma$ exists so we must have β of size $n \times r$ and γ of size $r \times s$. Then $\beta\gamma$ has size $n \times s$ so since $\alpha(\beta\gamma)$ exists, α must have size $m \times n$. With these sizes we see that the right hand side is also defined. Conversely if the right hand side exists, then so does the left and the matrix sizes must be as above.

Say $\alpha = \left[a_{ij}\right]$, $\beta = \left[b_{ij}\right]$, $\gamma = \left[c_{ij}\right]$. Let $\beta\gamma = \left[e_{ij}\right]$ and $\alpha(\beta\gamma) = \left[f_{ij}\right]$. Then

$$e_{ij} = \sum_{k=1}^{r} b_{ik}c_{kj}$$

and

$$f_{ij} = \sum_{k'=1}^{n} a_{ik'} e_{k'j} = \sum_{k'=1}^{n} a_{ik'} \sum_{k=1}^{r} b_{k'k} c_{kj}$$

$$= \sum_{k'=1}^{n} \sum_{k=1}^{r} a_{ik'} b_{k'k} c_{kj}$$

Similarly, if $(\alpha\beta)\gamma = \left[f_{ij}' \right]$ then

$$f_{ij}' = \sum_{k=1}^{r} \sum_{k'=1}^{n} a_{ik'} b_{k'k} c_{kj} = f_{ij}$$

and the associative law holds.

Now consider the equation $\alpha(\beta + \gamma) = \alpha\beta + \alpha\gamma$. Clearly the condition that makes either side defined is that α has size $m \times n$ and that both β and γ have size $n \times s$. Let $\alpha = \left[a_{ij} \right]$, $\beta = \left[b_{ij} \right]$, $\gamma = \left[c_{ij} \right]$ and $\alpha(\beta + \gamma) = \left[e_{ij} \right]$. Then

$$\left[e_{ij} \right] = \left[a_{ij} \right] \left(\left[b_{ij} \right] + \left[c_{ij} \right] \right) = \left[a_{ij} \right] \left[b_{ij} + c_{ij} \right]$$

so

$$e_{ij} = \sum_{k=1}^{n} a_{ik}(b_{kj} + c_{kj})$$

$$= \left(\sum_{k=1}^{n} a_{ik} b_{kj} \right) + \left(\sum_{k=1}^{n} a_{ik} c_{kj} \right)$$

Now the first term directly above is clearly the i,j-th entry in $\alpha\beta = \left[a_{ij} \right] \left[b_{ij} \right]$ and the second term is the i,j-th entry in $\alpha\gamma = \left[a_{ij} \right] \left[c_{ij} \right]$. Thus by definition of addition we have immediately $\alpha(\beta+\gamma) = \alpha\beta+\alpha\gamma$.

The other distributive law follows in a similar manner and the lemma is proved. $\square$

It will be necessary to develop a certain amount of manipulative skill in computing matrix products. If $\alpha = \left[a_{ij} \right]$ is an $m \times n$ matrix and $\beta = \left[b_{ij} \right]$ is $n \times r$, then $\alpha\beta = \left[c_{ij} \right]$ is $m \times r$ and we have to find all mr entries. Clearly the easiest way to scan through all these mr entries is to work column by column or row by row.

Suppose we consider the column by column method. Say we wish to find the jth column of $\alpha\beta$. As we have already observed, this depends upon all of α but only the jth column of β. So we start by physically (or mentally) lifting the jth column of β, turning it on its side and placing it above the matrix α, as indicated in the following diagram.

$$\alpha = \begin{matrix} b_{1j} & b_{2j} & \cdots & b_{nj} \\ \begin{bmatrix} a_{11} & a_{12} & \cdots & a_{1n} \\ a_{21} & a_{22} & \cdots & a_{2n} \\ \multicolumn{4}{c}{\cdots\cdots\cdots} \\ a_{m1} & a_{m2} & \cdots & a_{mn} \end{bmatrix} \end{matrix} \qquad \beta = \begin{bmatrix} b_{1j} \\ b_{2j} \\ \vdots \\ b_{nj} \end{bmatrix}$$

We then apply the b-row to each of the a-rows in turn, multiplying the corresponding b- and a-entries and summing these products. The sum we get from the ith row of α is then

$$a_{i1}b_{1j} + a_{i2}b_{2j} + \cdots + a_{in}b_{nj} = c_{ij}$$

Thus, in this way we have found the ith entry of the jth column of $\alpha\beta$, namely c_{ij}.

We can also proceed row by row. Here to find the ith row of the product, we take the ith row of α, turn it on its side and place it next to β.

$$\alpha = \begin{bmatrix} & & & \\ a_{i1} & a_{i2} & \cdots & a_{in} \\ & & & \end{bmatrix} \qquad \beta = \begin{matrix} a_{i1} \\ a_{i2} \\ \vdots \\ a_{in} \end{matrix} \begin{bmatrix} b_{11} & b_{12} & \cdots & b_{1r} \\ b_{21} & b_{22} & \cdots & b_{2r} \\ \multicolumn{4}{c}{\cdots\cdots\cdots} \\ b_{n1} & b_{n2} & \cdots & b_{nr} \end{bmatrix}$$

We then apply the a-column in turn to each of the b-columns, multiplying the corresponding a- and b-entries and summing these products. The sum we get from the jth column of β is then

$$a_{i1}b_{1j} + a_{i2}b_{2j} + \cdots + a_{in}b_{nj} = c_{ij}$$

Thus in this way we find the jth entry of the ith row of $\alpha\beta$.

Problems

11.1. Complete the proof of Theorem 11.1 that $\mathcal{L}(V, W)$ is a vector space.

Let V be a vector space over $\mathbb{R}$ with basis $\mathcal{A} = \{\alpha_1, \alpha_2\}$ and let W have basis $\mathcal{B} = \{\beta_1, \beta_2\}$. Define $\alpha_1' = \alpha_1 + 2\alpha_2$, $\alpha_2' = 2\alpha_1 + 3\alpha_2$ in V and $\beta_1' = \beta_1 + 2\beta_2$, $\beta_2' = \beta_1 + \beta_2$ in W so that $\mathcal{A}' = \{\alpha_1', \alpha_2'\}$ is also a basis for V and $\mathcal{B}' = \{\beta_1', \beta_2'\}$ is a second basis for W.

11.2. Let $T \colon V \to W$ be given by

$$_{\mathcal{A}}[T]_{\mathcal{B}} = \begin{bmatrix} 1 & 4 \\ 5 & 2 \end{bmatrix}$$

Find $_{\mathcal{A}'}[T]_{\mathcal{B}}$ and $_{\mathcal{A}'}[T]_{\mathcal{B}'}$.

11.3. Let $I_V \colon V \to V$ be the identity map. Find $_{\mathcal{A}}[I_V]_{\mathcal{A}}$, $_{\mathcal{A}'}[I_V]_{\mathcal{A}}$ and $_{\mathcal{A}}[I_V]_{\mathcal{A}'}$. Compute the matrix product $_{\mathcal{A}}[I_V]_{\mathcal{A}'} {}_{\mathcal{A}'}[I_V]_{\mathcal{A}}$.

11.4. If $I_W \colon W \to W$ is the identity map, find $_{\mathcal{B}}[I_W]_{\mathcal{B}'}$. Compute the matrix product $_{\mathcal{A}'}[I_V]_{\mathcal{A}} {}_{\mathcal{A}}[T]_{\mathcal{B}} {}_{\mathcal{B}}[I_W]_{\mathcal{B}'}$ and compare this to $_{\mathcal{A}'}[T]_{\mathcal{B}'}$.

Let α, β, γ be matrices over F.

11.5. When are both products $\alpha\beta$ and $\beta\alpha$ defined? When do the products have the same size?

11.6. Show that $\alpha(\beta+\gamma)$ is defined if and only if $\alpha\beta+\alpha\gamma$ is defined and find the common condition on the sizes of α, β, γ that guarantees this.

11.7. Use the techniques suggested at the end of this section to perform the matrix multiplication

$$\begin{bmatrix} 1 & -1 & 2 \\ 3 & 1 & 0 \\ 2 & -4 & 1 \end{bmatrix} \begin{bmatrix} 1 & 0 \\ -2 & 4 \\ 3 & -2 \end{bmatrix}$$

Is it more efficient to use the row by row method or the column by column method?

11.8. Let $R = F^{n \times n}$. Then R has defined on it an addition and a multiplication. Which of the field axioms does R satisfy? A set with this sort of arithmetic is known as a *ring*. In particular, R is the ring of $n \times n$ matrices.

11.9. Recall that for any $a \in F$ the linear transformation $T_a \colon V \to V$ is defined by $\alpha T_a = a\alpha$. If $\mathcal{A} = \{\alpha_1, \alpha_2, \ldots, \alpha_n\}$ is a basis for V, find the matrix $_{\mathcal{A}}[T_a]_{\mathcal{A}}$.

11.10. Let $T \colon V \to W$ with

$$_{\mathcal{A}}[T]_{\mathcal{B}} = \begin{bmatrix} 0 & 2 & -1 \\ 1 & 1 & -4 \\ -1 & 3 & 2 \end{bmatrix}$$

If $\mathcal{A} = \{\alpha_1, \alpha_2, \alpha_3\}$, find $\ker T$.

12. Products of Linear Transformations

Now we have seen that the addition of matrices corresponds to addition of linear transformations and we have defined a multiplication of matrices. What does this correspond to? Obviously to a multiplication of linear transformations.

Let $S\colon U \to V$ and $T\colon V \to W$ be linear transformations. Then we can define the (composition) product $ST\colon U \to W$ by

$$\alpha(ST) = (\alpha S)T$$

for all $\alpha \in U$. In other words, we first apply S to map U into V, and then we apply T to map $(U)S \subseteq V$ into W. We first observe that this product is indeed a linear transformation.

Let $\alpha, \beta \in U$ and $a \in F$. Then

$$
\begin{aligned}
(\alpha + \beta)(ST) &= ((\alpha + \beta)S)T && \text{by definition of } ST \\
&= (\alpha S + \beta S)T && \text{since } S \text{ is a linear} \\
&&& \text{transformation} \\
&= (\alpha S)T + (\beta S)T && \text{since } T \text{ is a linear} \\
&&& \text{transformation} \\
&= \alpha(ST) + \beta(ST) && \text{by definition of } ST
\end{aligned}
$$

and similarly

$$
\begin{aligned}
(a\alpha)(ST) &= ((a\alpha)S)T && \text{by definition of } ST \\
&= (a(\alpha S))T && \text{since } S \text{ is a linear} \\
&&& \text{transformation} \\
&= a((\alpha S)T) && \text{since } T \text{ is a linear} \\
&&& \text{transformation} \\
&= a(\alpha(ST)) && \text{by definition of } ST
\end{aligned}
$$

Thus ST is a linear transformation.

It is natural to ask whether this multiplication has nice properties and it does. Of course, there is no reason to expect it to be commutative since in fact ST may be defined while TS is not. However we do have

LEMMA 12.1. *Multiplication of linear transformations is associative and distributive. Furthermore, $a(ST) = (aS)T = S(aT)$ for all $a \in F$ and appropriate linear transformations S and T.*

PROOF. Let us first consider the associative law. Let $R\colon U \to V$, $S\colon V \to W$ and $T\colon W \to X$ be three linear transformations with

U, V, W, X being vector spaces over the same field F. We compare $R(ST)$ and $(RS)T$ which both map U to X. Let $\alpha \in U$. Then

$$
\begin{aligned}
\alpha(R(ST)) &= (\alpha R)(ST) && \text{by definition of } R(ST) \\
&= ((\alpha R)S)T && \text{by definition of } ST
\end{aligned}
$$

and

$$
\begin{aligned}
\alpha((RS)T) &= (\alpha(RS))T && \text{by definition of } (RS)T \\
&= ((\alpha R)S)T && \text{by definition of } RS
\end{aligned}
$$

Since these are equal for all $\alpha \in U$, the associative law holds. Observe that what we have shown above is that no matter how we write the product RST, the element $\alpha(RST)$ is found by first applying R, then S and then T.

Now suppose $R\colon U \to V$ and $S, T\colon V \to W$. Then for all $\alpha \in U$,

$$
\begin{aligned}
\alpha(R(S+T)) &= (\alpha R)(S+T) && \text{by definition of } R(S+T) \\
&= (\alpha R)S + (\alpha R)T && \text{by definition of } S+T \\
&= \alpha(RS) + \alpha(RT) && \text{by definition of } RS, RT \\
&= \alpha(RS + RT) && \text{by definition of } RS + RT
\end{aligned}
$$

Thus $R(S+T) = RS + RT$.

Next, let $S, T\colon U \to V$ and $R\colon V \to W$. Then for all $\alpha \in U$,

$$
\begin{aligned}
\alpha((S+T)R) &= (\alpha(S+T))R && \text{by definition of } (S+T)R \\
&= (\alpha S + \alpha T)R && \text{by definition of } S+T \\
&= (\alpha S)R + (\alpha T)R && \text{since } R \text{ is a linear} \\
& && \text{transformation} \\
&= \alpha(SR) + \alpha(TR) && \text{by definition of } SR, TR \\
&= \alpha(SR + TR) && \text{by definition of } SR + TR
\end{aligned}
$$

Thus $(S+T)R = SR + TR$ and the distributive laws are proved.

Finally, let $S\colon U \to V$, $T\colon V \to W$ and let $a \in F$. Then for all $\alpha \in U$, we have

$$
\begin{aligned}
\alpha(S(aT)) &= (\alpha S)(aT) && \text{by definition of } S(aT) \\
&= a((\alpha S)T) && \text{by definition of } aT \\
&= a(\alpha(ST)) && \text{by definition of } ST \\
&= \alpha(a(ST)) && \text{by definition of } a(ST)
\end{aligned}
$$

and

$$\alpha((aS)T) = (\alpha(aS))T \qquad \text{be definition of } (aS)T$$
$$= (a(\alpha S))T \qquad \text{by definition of } aS$$
$$= a((\alpha S)T) \qquad \text{since } T \text{ is linear}$$
$$= a(\alpha(ST)) \qquad \text{by definition of } ST$$
$$= \alpha(a(ST)) \qquad \text{by definition of } a(ST)$$

Thus $S(aT) = a(ST) = (aS)T$, as required. $\square$

Notice that the associative and first distributive law follow formally from the definition of product and sum of functions. But the second distributive law requires that R is a linear transformation. Similarly for the two formulas involving $a \in F$. One is formal, but one requires linearity. Our main result here is the correspondence between multiplication of matrices and of linear transformations.

THEOREM 12.1. *Let* $S\colon U \to V$ *and* $T\colon V \to W$ *be linear transformations with* U, V, W *finite dimensional vector spaces over the same field* F. *Let* $\mathcal{A}$ *be a basis for* U, $\mathcal{B}$ *a basis for* V, *and* $\mathcal{C}$ *a basis for* W. *Then*

$$_\mathcal{A}[S]_\mathcal{B} \cdot {_\mathcal{B}}[T]_\mathcal{C} = {_\mathcal{A}}[ST]_\mathcal{C}$$

PROOF. Let $\mathcal{A} = \{\alpha_1, \alpha_2, \ldots, \alpha_m\}$, $\mathcal{B} = \{\beta_1, \beta_2, \ldots, \beta_n\}$ and let $\mathcal{C} = \{\gamma_1, \gamma_2, \ldots, \gamma_r\}$. Then $_\mathcal{A}[S]_\mathcal{B} = [a_{ij}]$ is an $m \times n$ matrix, $_\mathcal{B}[T]_\mathcal{C} = [b_{ij}]$ is $n \times r$, so their product exists and has size $m \times r$, the same size as $_\mathcal{A}[ST]_\mathcal{C} = [c_{ij}]$.

Now by definition of this correspondence, we have

$$\alpha_i S = \sum_k a_{ik}\beta_k$$

and

$$\beta_k T = \sum_j b_{kj}\gamma_j$$

so

$$\alpha_i(ST) = (\alpha_i S)T = \left(\sum_k a_{ik}\beta_k\right)T$$

$$= \sum_k a_{ik}(\beta_k T) = \sum_k a_{ik} \sum_j b_{kj}\gamma_j$$

$$= \sum_j \left(\sum_k a_{ik}b_{kj}\right)\gamma_j$$

On the other hand, by the definition of $_A[ST]_C = [c_{ij}]$, we have

$$\alpha_i(ST) = \sum_j c_{ij}\gamma_j$$

This yields

$$c_{ij} = \sum_k a_{ik}b_{kj}$$

so $[c_{ij}] = [a_{ij}] \cdot [b_{ij}]$ and the result follows.   □

This simple result of course explains our definition of matrix multiplication. Indeed matrix multiplication was defined so that it would correspond to the composition of linear transformations.

The result also has numerous corollaries. Suppose $T : V \to W$ and T is given in the form of its corresponding matrix $_A[T]_B$ for some pair of bases $\mathcal{A}$ and $\mathcal{B}$. Now it is quite possible that if another pair $\mathcal{A}'$ and $\mathcal{B}'$ were chosen, then the matrix $_{A'}[T]_{B'}$ would be so simple that we could easily visualize the action of T. We will see an example of this later. A natural problem is then to find the relationship between $_A[T]_B$ and $_{A'}[T]_{B'}$. Recall that matrix multiplication is associative, so a product of matrices can be defined unambiguously without the use of parentheses.

COROLLARY 12.1. *Let $T : V \to W$ be a linear transformation with V and W finite dimensional vector spaces over F. Let $\mathcal{A}, \mathcal{A}'$ be bases of V and let $\mathcal{B}, \mathcal{B}'$ be bases for W. Then*

$$_{A'}[T]_{B'} = {_{A'}[I_V]_A} \cdot {_A[T]_B} \cdot {_B[I_W]_{B'}}$$

where $I_V : V \to V$ and $I_W : W \to W$ are the identity maps.

PROOF. By the previous theorem

$$_{A'}[I_V]_A \cdot \left({_A[T]_B} \cdot {_B[I_W]_{B'}}\right) = {_{A'}[I_V]_A} \cdot {_A[T \cdot I_W]_{B'}}$$
$$= {_{A'}[I_V \cdot T \cdot I_W]_{B'}}$$

Since clearly $T = I_V T I_W$, the result follows.   ⊓

For obvious reasons, the matrices $_{A'}[I_V]_A$ and $_B[I_W]_{B'}$ are called the *change of basis* matrices. They depend upon $\mathcal{A}, \mathcal{A}', \mathcal{B}$ and $\mathcal{B}'$, but not on the linear transformation T itself. Of course, we are now faced with determining the nature of these matrices. Just how special are they?

In studying these change of basis matrices, we see first of all that they correspond to linear transformations from a space to itself. So for the time being, we will restrict all linear transformations to be from V to V. Let $\mathcal{A} = \{\alpha_1, \alpha_2, \ldots, \alpha_n\}$ and $\mathcal{A}' = \{\alpha_1', \alpha_2', \ldots, \alpha_n'\}$ be bases for

V and let $I \colon V \to V$ be the identity transformation. If $_{\mathcal{A}'}[I]_{\mathcal{A}} = [a_{ij}]$, then

$$\alpha_i' = \alpha_i' I = a_{i1}\alpha_1 + a_{i2}\alpha_2 + \cdots + a_{in}\alpha_n$$

In other words, the entries in the ith row of $[a_{ij}]$ are merely the coefficients that occur when we write α_i' in terms of the basis $\mathcal{A}$. If $\mathcal{A} \neq \mathcal{A}'$ then of course $[a_{ij}]$ might be quite complicated. On the other hand if $\mathcal{A} = \mathcal{A}'$ (in the same order) then clearly $_{\mathcal{A}}[I]_{\mathcal{A}} = I_n$ is the $n \times n$ *identity matrix*

$$I_n = \begin{bmatrix} 1 & & & \\ & 1 & & 0 \\ & & 1 & \\ & 0 & & \ddots \\ & & & & 1 \end{bmatrix}$$

That is $I_n = [\delta_{ij}]$ where

$$\delta_{ij} = \begin{cases} 1, & \text{for } i = j \\ 0, & \text{for } i \neq j \end{cases}$$

Certainly $_{\mathcal{A}}[T]_{\mathcal{A}} = I_n$ if and only if $T = I$. Thus we can easily identify the identity transformation from the matrix $_{\mathcal{A}}[I]_{\mathcal{A}}$, but not so easily from a random $_{\mathcal{A}'}[I]_{\mathcal{A}}$.

Now I_n is easily seen to be the identity element in $F^{n \times n}$, that is $I_n \alpha = \alpha I_n = \alpha$ for all $\alpha \in F^{n \times n}$. Next let $\beta \in F^{n \times n}$. We say that β is *nonsingular* if and only if β has an inverse $\beta^{-1} \in F^{n \times n}$ with $\beta \beta^{-1} = \beta^{-1}\beta = I_n$.

LEMMA 12.2. *Let $\beta = {_{\mathcal{A}'}[T]_{\mathcal{A}}}$ with $T \colon V \to V$. Then β is nonsingular if and only if T is nonsingular and when this occurs we have $\beta^{-1} = {_{\mathcal{A}}[T^{-1}]_{\mathcal{A}'}}$.*

PROOF. Suppose T is nonsingular. Then T is one-to-one and onto and hence T^{-1} exists. From the definition of T^{-1} we have clearly $TT^{-1} = T^{-1}T = I$. Thus

$$_{\mathcal{A}'}[T]_{\mathcal{A}} \cdot {_{\mathcal{A}}[T^{-1}]_{\mathcal{A}'}} = {_{\mathcal{A}'}[TT^{-1}]_{\mathcal{A}'}} = {_{\mathcal{A}'}[I]_{\mathcal{A}'}} = I_n$$

and

$$_{\mathcal{A}}[T^{-1}]_{\mathcal{A}'} \cdot {_{\mathcal{A}'}[T]_{\mathcal{A}}} = {_{\mathcal{A}}[T^{-1}T]_{\mathcal{A}}} = {_{\mathcal{A}}[I]_{\mathcal{A}}} = I_n$$

Therefore β has an inverse, namely $_{\mathcal{A}}[T^{-1}]_{\mathcal{A}'}$.

Conversely, suppose β is nonsingular and, by Theorem 11.2, let $S \colon V \to V$ be given by $\beta^{-1} = {_{\mathcal{A}}[S]_{\mathcal{A}'}}$. Then

$$_{\mathcal{A}}[ST]_{\mathcal{A}} = {_{\mathcal{A}}[S]_{\mathcal{A}'}} \cdot {_{\mathcal{A}'}[T]_{\mathcal{A}}} = \beta^{-1}\beta = I_n$$

so $ST = I$. Similarly

$$_{A'}[TS]_{A'} = {}_{A'}[T]_A \cdot {}_A[S]_{A'} = \beta\beta^{-1} = I_n$$

so $TS = I$. Observe that for $\alpha \in V$, $(\alpha S)T = \alpha I = \alpha$ so T is onto. Also $(\alpha T)S = \alpha I = \alpha$, so $\alpha T = 0$ implies $\alpha = 0$ and T is one-to-one. Thus T is nonsingular and then clearly $S = T^{-1}$, so $\beta^{-1} = {}_A[T^{-1}]_{A'}$ and the result follows. □

In particular, since I is nonsingular with inverse $I^{-1} = I$, the above shows that every change of basis matrix $\beta = {}_{A'}[I]_A$ is nonsingular with inverse $\beta^{-1} = {}_A[I]_{A'}$. Now for the converse.

LEMMA 12.3. *Let β be a nonsingular matrix and let A be a basis for V. Then there exist bases A' and A'' of V with*

$$\beta = {}_A[I]_{A'} = {}_{A''}[I]_A$$

In particular, a square matrix is a change of basis matrix if and only if it is nonsingular.

PROOF. Choose $T\colon V \to V$ so that $\beta = {}_A[T]_A$. Since β is nonsingular, so is T. Let $A'' = (A)T$ be the image of A under T and let $A' = (A)T^{-1}$. Since A spans V, it is clear that A'' spans $\operatorname{im} T = V$, and A' spans $\operatorname{im} T^{-1} = V$. Since these are spanning sets of size $n = \dim_F V$, they are therefore bases for V. Now clearly

$$I_n = {}_A[I]_{A''} = {}_A[T^{-1}]_{A'}$$

so

$$\beta = {}_A[T]_A \cdot I_n = {}_A[T]_A \cdot {}_A[T^{-1}]_{A'} = {}_A[I]_{A'}$$

and

$$\beta = I_n^{-1}\beta = {}_A[T]_{A''}^{-1} \cdot {}_A[T]_A$$
$$= {}_{A''}[T^{-1}]_A \cdot {}_A[T]_A = {}_{A''}[I]_A$$

The lemma is proved. □

Let U, V and W be vector spaces over F and suppose $T\colon V \to W$ is a linear transformation. Then via composition, T induces map $T_1\colon \mathcal{L}(W, U) \to \mathcal{L}(V, U)$ and $T_2\colon \mathcal{L}(U, V) \to \mathcal{L}(U, W)$ given by

$$T_1\colon S \mapsto TS \qquad \text{and} \qquad T_2\colon S \mapsto ST$$

It follows easily from Lemma 12.1 that both T_1 and T_2 are linear transformations. We will use this in the special case below where U is the 1-dimensional vector space F.

Let V be a vector space over F. Then $\mathcal{L}(V, F)$ is called the *dual space* of V and is denoted by V^*. Furthermore, the elements $\lambda \in V^*$

are called *linear functionals*. Of course, these are just linear transformations so that $(\alpha_1 + \alpha_2)\lambda = \alpha_1\lambda + \alpha_2\lambda$ and $(c\alpha_1)\lambda = c(\alpha_1\lambda)$ for all $\alpha_1, \alpha_2 \in V$ and $c \in F$.

LEMMA 12.4. *Let* $\dim_F V = n$ *and let* $\mathcal{A} = \{\alpha_1, \alpha_2, \ldots, \alpha_n\}$ *be a basis for* V. *Then the functionals* $\alpha_i^*\colon V \to F$ *defined by*

$$\alpha_i^*\colon \alpha_j \mapsto \begin{cases} 1, & \text{if } j = i \\ 0, & \text{if } j \neq i \end{cases}$$

form a basis $\mathcal{A}^* = \{\alpha_1^*, \alpha_2^*, \ldots, \alpha_n^*\}$, *the dual basis, for* V^*. *In particular,* $\dim_F V^* = \dim_F V$.

PROOF. Since $\mathcal{A}$ is a basis for V, Theorem 8.1 implies that there exists one and only one linear functional $\beta\colon V \to F$ with $(\alpha_i)\beta = a_i$ for each choice of $a_1, a_2, \ldots, a_n \in F$. In particular, it follows that the dual functionals α_i^* exist. Furthermore, since $\beta = a_1\alpha_1^* + a_2\alpha_2^* + \cdots + a_n\alpha_n^*$ maps α_i to a_i for all i, we see that this β must be the unique functional sending α_i to a_i. We now conclude easily that $\mathcal{A}^*$ is a basis for V^*. $\square$

Recall that if $\alpha = [a_{ij}]$ is the $m \times n$ matrix given by

$$\alpha = \begin{bmatrix} a_{11} & a_{12} & \cdots & a_{1n} \\ a_{21} & a_{22} & \cdots & a_{2n} \\ \multicolumn{4}{c}{\cdots\cdots\cdots} \\ a_{m1} & a_{m2} & \cdots & a_{mn} \end{bmatrix} \in F^{m \times n}$$

then the *transpose* α^T of α is defined to be the $n \times m$ matrix

$$\alpha^\mathsf{T} = \begin{bmatrix} a_{11} & a_{21} & \cdots & a_{m1} \\ a_{12} & a_{22} & \cdots & a_{m2} \\ \multicolumn{4}{c}{\cdots\cdots\cdots} \\ a_{1n} & a_{2n} & \cdots & a_{mn} \end{bmatrix} \in F^{n \times m}$$

In other words, $\alpha^\mathsf{T} = [a'_{ij}]$ where $a'_{ij} = a_{ji}$ for all appropriate i and j.

THEOREM 12.2. *Let* V *and* W *be finite dimensional* F-*vector spaces with bases* $\mathcal{A}$ *and* $\mathcal{B}$ *respectively, and let* $T\colon V \to W$ *be a linear transformation. Then the map* $T^*\colon W^* \to V^*$ *given by* $\lambda \mapsto T\lambda$ *for all* $\lambda \in W^*$ *is a linear transformation and the corresponding matrices* $_\mathcal{A}[T]_\mathcal{B}$ *and* $_{\mathcal{B}^*}[T^*]_{\mathcal{A}^*}$ *are transposes of each other.*

PROOF. We have already observed that $T^*\colon W^* \to V^*$ is a linear transformation. Now, let $_\mathcal{A}[T]_\mathcal{B} = [a_{ij}]$ so that $\alpha_i T = \sum_j a_{ij}\beta_j$ and let us write $_{\mathcal{B}^*}[T^*]_{\mathcal{A}^*} = [a_{ji}^*]$ so that $\beta_j^* T^* = \sum_i a_{ji}^* \alpha_i^*$. Then $\beta_j^* T^*$ is the unique functional in V^* that sends α_i to a_{ji}^*. Furthermore, by definition, $\beta_j^* T^*$ is the composite map $T\beta_j^*$, so this functional sends α_i

to $(\alpha_i T)\beta_j^* = (\sum_k a_{ik}\beta_k)\beta_j^* = a_{ij}$. Thus $a_{ji}^* = a_{ij}$ and the matrices $_\mathcal{A}[T]_\mathcal{B}$ and $_{\mathcal{B}^*}[T^*]_{\mathcal{A}^*}$ are indeed transposes of each other.   □

Problems

12.1. Suppose $T_1\colon V_1 \to V_2$, $T_2\colon V_2 \to V_3$, ..., $T_k\colon V_k \to V_{k+1}$ are linear transformations or in fact any functions. Show that the composition product $T_1 T_2 \cdots T_k$ with any choice of parentheses merely amounts to first applying T_1, then T_2, ..., and then T_k.

12.2. Let $S\colon U \to V$ and $T\colon V \to W$ be linear transformations with U, V and W finite dimensional. Prove that

$$\min\{\operatorname{rank} S, \operatorname{rank} T\} \geq \operatorname{rank} ST \geq \operatorname{rank} S + \operatorname{rank} T - \dim V$$

For the second inequality, let R be the restriction of T to $\operatorname{im} S \subseteq V$. Then $\operatorname{im} R = \operatorname{im} ST$ and $\ker R = \operatorname{im} S \cap \ker T \subseteq \ker T$, so

$$\dim \ker R \leq \dim \ker T = \dim V - \operatorname{rank} T$$

12.3. Let $\alpha \in F^{m\times n}$ and $\beta \in F^{n\times r}$ so that $\alpha\beta \in F^{m\times r}$. Use Theorem 11.2 and the preceding problem with $U = F^m$, $V = F^n$ and $W = F^r$ to show that

$$\min\{\operatorname{rank} \alpha, \operatorname{rank} \beta\} \geq \operatorname{rank} \alpha\beta \geq \operatorname{rank} \alpha + \operatorname{rank} \beta - n$$

12.4. Prove that I_n is the unique identity element of the ring $F^{n\times n}$. More generally, if $\alpha \in F^{m\times n}$ and $\beta \in F^{n\times k}$ show that $\alpha I_n = \alpha$ and $I_n\beta = \beta$.

12.5. Show how to deduce the associative and distributive laws of matrix multiplication from those of linear transformation multiplication.

12.6. Let $T\colon V \to W$ be a linear transformation with $_\mathcal{A}[T]_\mathcal{B} = [c_{ij}]$ where $\mathcal{A} = \{\alpha_1, \alpha_2, \ldots, \alpha_m\}$ and $\mathcal{B} = \{\beta_1, \beta_2, \ldots, \beta_n\}$ are bases. Let

$$\alpha = a_1\alpha_1 + a_2\alpha_2 + \cdots + a_m\alpha_m \in V$$

and

$$\alpha T = \beta = b_1\beta_1 + b_2\beta_2 + \cdots + b_n\beta_n \in W$$

Show that matrix multiplication yields

$$\begin{bmatrix} a_1 & a_2 & \cdots & a_m \end{bmatrix} \begin{bmatrix} c_{11} & c_{12} & \cdots & c_{1n} \\ c_{21} & c_{22} & \cdots & c_{2n} \\ \multicolumn{4}{c}{\cdots\cdots\cdots} \\ c_{m1} & c_{m2} & \cdots & c_{mn} \end{bmatrix} = \begin{bmatrix} b_1 & b_2 & \cdots & b_n \end{bmatrix}$$

How does this relate to Theorem 12.1?

12.7. Find the rank of the linear transformation T if

$$_A[T]_B = \begin{bmatrix} 1 & 0 & 3 & -2 \\ -1 & 2 & 1 & 4 \\ -1 & 6 & 9 & 8 \end{bmatrix}$$

12.8. Use Lemma 12.1 to prove in detail that the maps T_1 and T_2 described after Lemma 12.3 are both linear transformations.

12.9. If α and β are matrices of appropriate sizes so that $\alpha\beta$ exists, prove that $(\alpha\beta)^\mathsf{T} = \beta^\mathsf{T}\alpha^\mathsf{T}$.

12.10. Let $\beta \in F^{n \times n}$. Show that β is nonsingular if and only if β^T is nonsingular.

13. Eigenvalues and Eigenvectors

Suppose $T\colon V \to W$ is a linear transformation. What does it look like? To start with, by Theorem 9.3, there exist bases $\mathcal{A}$ and $\mathcal{B}$ of V and W respectively with

$$_\mathcal{A}[T]_\mathcal{B} = D_r = [\delta_{ij}]$$

where $\delta_{11} = \delta_{22} = \cdots = \delta_{rr} = 1$ and all remaining δ_{ij} are 0. Here, of course, r is the rank of T. In this way, T is clearly described in the nicest possible manner. But what happens when $W = V$? Does it really make sense to choose two different bases for V in order to describe T? The answer is definitely "no". For example. suppose we are given $\mathcal{A}$ and $\mathcal{B}$, both bases for V, with $_\mathcal{A}[T]_\mathcal{B} = D_r$. Then we will obviously want to find some relationship between $\mathcal{A}$ and $\mathcal{B}$, and this clearly reduces the situation to the case of a single basis. Thus our main problem here is to find a basis $\mathcal{A}$ for V so that $_\mathcal{A}[T]_\mathcal{A}$ is as simple as possible.

For the remainder of this section, V will be a fixed vector space over F of finite dimension and all linear transformations will map V to V. Before we concern ourselves with the problem of choosing a nice basis for T, let us consider what happens to the matrix associated with T under any sort of change of basis.

LEMMA 13.1. *Let $T\colon V \to V$ and let $\mathcal{A}$ be a basis for V.*

    i. If $\mathcal{B}$ is a second basis for V, then

$$_\mathcal{B}[T]_\mathcal{B} = \beta^{-1} \cdot {}_\mathcal{A}[T]_\mathcal{A} \cdot \beta$$

    where β is the nonsingular matrix $_\mathcal{A}[I]_\mathcal{B}$.

    ii. Conversely, given a nonsingular matrix $\beta \in F^{n \times n}$ there exists a basis $\mathcal{B}$ such that $_\mathcal{B}[T]_\mathcal{B}$ is given as above.

PROOF. We know that

$$_\mathcal{B}[T]_\mathcal{B} = {}_\mathcal{B}[I]_\mathcal{A} \cdot {}_\mathcal{A}[T]_\mathcal{A} \cdot {}_\mathcal{A}[I]_\mathcal{B}$$

Moreover if $\beta = {}_\mathcal{A}[I]_\mathcal{B}$, then $\beta^{-1} = {}_\mathcal{B}[I]_\mathcal{A}$ so

$$_\mathcal{B}[T]_\mathcal{B} = \beta^{-1} \cdot {}_\mathcal{A}[T]_\mathcal{A} \cdot \beta$$

and this yields (i).

Conversely, if β is any nonsingular matrix, then Lemma 12.3 implies that there exists a basis $\mathcal{B}$ with $\beta = {}_\mathcal{A}[I]_\mathcal{B}$. Since $\beta^{-1} = {}_\mathcal{B}[I]_\mathcal{A}$, it then follows that

$$\beta^{-1} \cdot {}_\mathcal{A}[T]_\mathcal{A} \cdot \beta = {}_\mathcal{B}[T]_\mathcal{B}$$

and the lemma is proved. $\qquad\square$

Let $\alpha, \gamma \in F^{n \times n}$. We say that α and γ are *similar* if there exists a nonsingular matrix β with $\gamma = \beta^{-1} \alpha \beta$. Thus the above lemma says that α and γ are similar if and only if they represent the same linear transformation but over possibly different bases, that is $\alpha = {}_{\mathcal{A}}[T]_{\mathcal{A}}$ and $\gamma = {}_{\mathcal{B}}[T]_{\mathcal{B}}$.

Now let us consider some examples to see how nice we can hope ${}_{\mathcal{A}}[T]_{\mathcal{A}}$ to be.

EXAMPLE 13.1. If $I \colon V \to V$ is the identity map, then certainly for all bases $\mathcal{A}$ we have ${}_{\mathcal{A}}[I]_{\mathcal{A}} = I_n$ where $\dim_F V = n$.

EXAMPLE 13.2. Let $T_a \colon V \to V$ denote scalar multiplication by $a \in F$. Then for any $\alpha \in V$ we have $\alpha T_a = a\alpha$, so clearly

$$
{}_{\mathcal{A}}[T_a]_{\mathcal{A}} = \begin{bmatrix} a & & & \\ & a & & 0 \\ & & \ddots & \\ 0 & & & a \end{bmatrix} = aI_n
$$

Thus all entries down the *main diagonal* (namely the i, i-entries for $i = 1, 2, \ldots, n$) are equal to a and the remaining terms are 0.

EXAMPLE 13.3. The above shows that we cannot hope to always get matrices of the form D_r. In fact, what happens when ${}_{\mathcal{A}}[T]_{\mathcal{A}} = D_r$? Say $\mathcal{A} = \{\alpha_1, \alpha_2, \ldots, \alpha_n\}$ and let $W = \langle \alpha_1, \alpha_2, \ldots, \alpha_r \rangle$ and $W' = \langle \alpha_{r+1}, \alpha_{r+2}, \ldots, \alpha_n \rangle$. Then certainly $V = W \oplus W'$, $\alpha T = \alpha$ for $\alpha \in W$, and $\alpha T = 0$ for $\alpha \in W'$. This shows that T is essentially the projection from V to W followed by the embedding of W back into V. In other words, if $\alpha \in V$ satisfies $\alpha = \beta + \beta'$ with $\beta \in W$ and $\beta' \in W'$, then $\alpha T = \beta$ where we think of β as being an element of V.

These three examples all have one thing in common. The matrices ${}_{\mathcal{A}}[T]_{\mathcal{A}}$ above are all *diagonal matrices*, that is their only nonzero entries occur on the main diagonal. Since, in such matrices, we need only be concerned with the n diagonal entries, we can use the shorthand notation $\operatorname{diag}(a_1, a_2, \ldots, a_n)$ for the $n \times n$ diagonal matrix

$$
\operatorname{diag}(a_1, a_2, \ldots, a_n) = \begin{bmatrix} a_1 & & & \\ & a_2 & & 0 \\ & & \ddots & \\ 0 & & & a_n \end{bmatrix}
$$

Let $T \colon V \to V$. If there exists a basis $\mathcal{A}$ such that ${}_{\mathcal{A}}[T]_{\mathcal{A}}$ is diagonal, then we say that T can be *diagonalized*.

Suppose ${}_{\mathcal{A}}[T]_{\mathcal{A}} = \operatorname{diag}(a_1, a_2, \ldots, a_n)$ where $\mathcal{A} = \{\alpha_1, \alpha_2, \ldots, \alpha_n\}$. Then by definition, we have $\alpha_i T = a_i \alpha_i$ for $i = 1, 2, \ldots, n$. In other words, T sends each α_i to a scalar multiple of itself. Vectors with this property are therefore of interest to us and we give them a name. Suppose $\alpha \in V, \alpha \neq 0$ and $\alpha T = a\alpha$ for some $a \in F$. Then we say that α is an *eigenvector* for T and a is its associated *eigenvalue*. The following theorem is now obvious. It is true by definition.

THEOREM 13.1. *Let $T \colon V \to V$ be a linear transformation. Then T can be diagonalized if and only if V has a basis consisting entirely of eigenvalues of T.*

Now we ask, can a linear transformation always be diagonalized? By the above we are asking for a full basis consisting of eigenvectors. But do eigenvectors necessarily exist and how do we find them? The answer as we will see in due time is that eigenvectors always exist if the field F is in some sense big enough, but even then there may not be enough to yield a basis for V. Let us consider some examples.

EXAMPLE 13.4. Let V be a 2-dimensional space over $\mathbb{R}$ with basis $\{\alpha_1, \alpha_2\}$ and suppose T is given by

$$\begin{aligned} \alpha_1 T &= \alpha_1 + 2\alpha_2 \\ \alpha_2 T &= -\alpha_1 - \alpha_2 \end{aligned}$$

Let $\alpha = a_1\alpha_1 + a_2\alpha_2 \in V$ with $\alpha T = c\alpha$ for some $c \in \mathbb{R}$. Then

$$\begin{aligned} ca_1\alpha_1 + ca_2\alpha_2 = c\alpha &= \alpha T \\ &= a_1(\alpha_1 + 2\alpha_2) + a_2(-\alpha_1 - \alpha_2) \\ &= (a_1 - a_2)\alpha_1 + (2a_1 - a_2)\alpha_2 \end{aligned}$$

This yields

$$ca_1 = a_1 - a_2, \qquad ca_2 = 2a_1 - a_2$$

so

$$(1 - c)a_1 = a_2, \qquad (1 + c)a_2 = 2a_1$$

Thus if we assume that $\alpha \neq 0$, then we have $a_1 \neq 0$, $a_2 \neq 0$ and

$$\frac{a_1}{a_2} = \frac{1}{1 - c} = \frac{1 + c}{2}$$

But this yields

$$0 = 2 - (1 + c)(1 - c) = 1 + c^2$$

and there is no such $c \in \mathbb{R}$ satisfying this equation. On the other hand $c = \sqrt{-1} \in \mathbb{C}$ is a solution, so the trouble here is that $\mathbb{R}$ is just not big enough.

EXAMPLE 13.5. Let V be as above, but now let T be given by

$$\alpha_1 T = - \alpha_1 + 2\alpha_2$$
$$\alpha_2 T = -2\alpha_1 + 3\alpha_2$$

Suppose $\alpha = a_1 \alpha_1 + a_2 \alpha_2 \in V$ is an eigenvector for T with corresponding eigenvalue $c \in \mathbb{R}$. Then

$$ca_1 \alpha_1 + ca_2 \alpha_2 = c\alpha = \alpha T$$
$$= a_1(-\alpha_1 + 2\alpha_2) + a_2(-2\alpha_1 + 3\alpha_2)$$
$$= (-a_1 - 2a_2)\alpha_1 + (2a_1 + 3a_2)\alpha_2$$

Thus

$$ca_1 = -a_1 - 2a_2, \qquad ca_2 = 2a_1 + 3a_2$$

so

$$(c+1)a_1 = -2a_2, \qquad (c-3)a_2 = 2a_1$$

Again if $\alpha \neq 0$, then $a_1 \neq 0$, $a_2 \neq 0$ and

$$\frac{a_1}{a_2} = \frac{-2}{c+1} = \frac{c-3}{2}$$

so

$$0 = (c+1)(c-3) + 4 = (c-1)^2$$

Thus $c = 1$, $a_2 = -a_1$ and $\alpha = a_1(\alpha_1 - \alpha_2)$. We have therefore found an eigenvector, namely $\alpha_1 - \alpha_2$, but all other eigenvectors are just scalar multiples of this one. Hence T cannot be diagonalized.

Suppose $T: V \to V$ and we wish to solve the equation $\alpha T = a\alpha$ for $\alpha \neq 0$. If a is known then it is apparent that this is merely a set of linear equations in the coefficients of α with respect to some basis of V and this offers no great difficulty. Thus the real problem is to find the eigenvalues. If $S: V \to V$ is a linear transformation, we say that S is *singular* if and only if it is not nonsingular. In view of Corollary 9.3 this occurs if and only if $\ker S \neq 0$. The following is a trivial reformulation of the eigenvalue problem.

THEOREM 13.2. *Let $T: V \to V$ be a linear transformation. Then $a \in F$ is an eigenvalue for T if and only if the transformation $S = aI - T$ is singular. When this occurs then $0 \neq \alpha \in \ker S$ if and only if α is an eigenvector of T with corresponding eigenvalue a.*

PROOF. $S = aI - T$ is singular if and only if there exists $\alpha \in V$, $\alpha \neq 0$ with

$$0 = \alpha S = \alpha(aI - T) = a\alpha - \alpha T$$

so the result is clear.   □

Let us try another example, a really ugly one this time.

EXAMPLE 13.6. Let V be a 3-dimensional vector space over $\mathbb{R}$ with basis $\{\alpha_1, \alpha_2, \alpha_3\}$ and suppose $T\colon V \to V$ is given by

$$\alpha_1 T = -10\alpha_1 + 9\alpha_2 + 6\alpha_3$$
$$\alpha_2 T = 32\alpha_1 - 23\alpha_2 - 18\alpha_3$$
$$\alpha_3 T = -61\alpha_1 + 47\alpha_2 + 35\alpha_3$$

If $a \in \mathbb{R}$ is an eigenvalue of T, then $S = aI - T$ is singular. Clearly

$$\alpha_1 S = (a + 10)\alpha_1 - 9\alpha_2 - 6\alpha_3$$
$$\alpha_2 S = -32\alpha_1 + (a + 23)\alpha_2 + 18\alpha_3$$
$$\alpha_3 S = 61\alpha_1 - 47\alpha_2 + (a - 35)\alpha_3$$

Now how do we determine when S is singular? By Corollary 9.3 this occurs if and only if $\ker S \neq 0$, so choose $\beta = b_1\alpha_1 + b_2\alpha_2 + b_3\alpha_3$ and suppose $\beta S = 0$. Indeed, in view of the preceding theorem, any such nonzero β will be an eigenvector of T. We have

$$\begin{aligned}
0 = \beta S &= b_1(\alpha_1 S) + b_2(\alpha_2 S) + b_3(\alpha_3 S) \\
&= ((a + 10)b_1 - 32b_2 + 61b_3)\alpha_1 \\
&\quad + (-9b_1 + (a + 23)b_2 - 47b_3)\alpha_2 \\
&\quad + (-6b_1 + 18b_2 + (a - 35)b_3)\alpha_3
\end{aligned}$$

so we must have

$$0 = h_1 - (a + 10)b_1 - 32b_2 + 61b_3$$
$$0 = h_2 = -9b_1 + (a + 23)b_2 - 47b_3$$
$$0 = h_3 = -6b_1 + 18b_2 + (a - 35)b_3$$

We solve this system of equations in b_1, b_2 and b_3 by eliminating one variable at a time. Since we do not know the value of $a \in \mathbb{R}$, we will therefore avoid dividing by any coefficients that depend upon it. Thus we start by saving the third equation and eliminating b_1 from the first and second. We get

$$0 = k_1 = h_3 \qquad\qquad = -6b_1 + 18b_2 + (a - 35)b_3$$
$$0 = k_2 = (a + 10)h_3 + 6h_1 \quad = (18a - 12)b_2 + (a^2 - 25a + 16)b_3$$
$$0 = k_3 = 6h_2 - 9h_3 \qquad = (6a - 24)b_2 + (33 - 9a)b_3$$

Unfortunately, at this point the coefficients in k_2 and k_3 all depend upon a and therefore the elimination procedure hits a slight snag. However, if we replace k_2 by $k_2 - 3k_3$ a simplification occurs

$$
\begin{aligned}
0 = \ell_1 = k_1 \qquad\qquad &= -6b_1 + 18b_2 + (a-35)b_3 \\
0 = \ell_2 = k_2 - 3k_3 \qquad\qquad &= 60b_2 + (a^2 + 2a - 83)b_3 \\
0 = \ell_3 = k_3 \qquad\qquad &= (6a - 24)b_2 + (33 - 9a)b_3
\end{aligned}
$$

Finally, we save the first two equations and eliminate b_2 from the third to obtain

$$
\begin{aligned}
0 = m_1 = \ell_1 \qquad\qquad &= -6b_1 + 18b_2 + (a-35)b_3 \\
0 = m_2 = \ell_2 \qquad\qquad &= 60b_2 + (a^2 + 2a - 83)b_3 \\
0 = m_3 = (a-4)\ell_2 - 10\ell_3 \quad &= (a^3 - 2a^2 - a + 2)b_3
\end{aligned}
$$

Suppose $(a^3 - 2a^2 - a + 2) \neq 0$. Then $0 = m_3$ implies that $b_3 = 0$ and we get in turn $b_2 = 0$ from $0 = m_2$ and $b_1 = 0$ from $0 = m_1$. Thus for a nonzero solution in the bs we must have

$$
0 = (a^3 - 2a^2 - a + 2) = (a-1)(a+1)(a-2)
$$

so $a = 1, -1$, or 2. For each of these three values it is now a simple matter to find nonzero b_1, b_2, b_3 and we get

$$
\begin{aligned}
a_1 = a = 1 \qquad\qquad &\beta_1 = \beta = -5\alpha_1 + 4\alpha_2 + 3\alpha_3 \\
a_2 = a = -1 \qquad\qquad &\beta_2 = \beta = -9\alpha_1 + 7\alpha_2 + 5\alpha_3 \\
a_3 = a = 2 \qquad\qquad &\beta_3 = \beta = -7\alpha_1 + 5\alpha_2 + 4\alpha_3
\end{aligned}
$$

Since $\beta S = 0$, it follows from Theorem 13.2 that β_i is an eigenvector for T with corresponding eigenvalue a_i. Now it is an easy matter to check (and in fact a consequence of a later theorem) that β_1, β_2 and β_3 are linearly independent. Thus V has a basis $\mathcal{B} = \{\beta_1, \beta_2, \beta_3\}$ consisting of eigenvectors of T, so

$$
{}_{\mathcal{B}}[T]_{\mathcal{B}} = \operatorname{diag}(1, -1, 2)
$$

and T is diagonalized. Certainly, T is now easily understood. We observe that here as in all earlier examples the eigenvalues of T occur as roots of certain polynomial equations.

Now it seems reasonable that the elimination method will probably work in general. But it is certainly unpleasant. There are difficulties with coefficients being functions of the possible eigenvalue and there are also difficulties with zero coefficients that we did not encounter here. Moreover there are certainly many different possibilities for the order in which we carry out this procedure so that the answer we get is not a priori strictly a function of T. Wouldn't it be nice therefore if there existed a fixed function of the entries of the matrix $_{\mathcal{A}}[S]_{\mathcal{A}}$ which would tell us at a glance just when S is singular. As we will see in the next chapter, there is a natural candidate for such a function and it indeed does the job.

Problems

13.1. For $\alpha, \beta \in F^{n \times n}$, let us write $\alpha \sim \beta$ if and only if α and β are similar. Prove that $\sim$ is an equivalence relation.

Let V be a fixed finite dimensional vector space over F and let $T \colon V \to V$. For each $a \in F$ we set

$$V_a = \{\alpha \in V \mid \alpha T = a\alpha\}$$

13.2. Show that V_a is a subspace of V.

13.3. Suppose $S \colon V \to V$ commutes with T, so that $ST = TS$. Prove that $(V_a)S \subseteq V_a$.

13.4. If a and b are distinct elements of F, show that $V_a \cap V_b = 0$ and conclude that $V_a + V_b = V_a \oplus V_b$.

13.5. Suppose T satisfies $T^2 = T$. Show that $V = V_0 \oplus V_1$ and that T is the projection map, with respect to this direct sum, of V into $V_1 \subseteq V$.

13.6. Suppose that every vector $\alpha \in V$ satisfies $\alpha T = a\alpha$ for some $a \in F$ depending upon α. Show that $V = V_b$ for some $b \in F$ so that $T = T_b$.

13.7. Assume that $V = V_{a_1} \oplus V_{a_2} \oplus \cdots \oplus V_{a_r}$ for suitable distinct $a_1, a_2, \ldots, a_r \in F$. Prove that T can be diagonalized.

13.8. Let $f(x) = \sum_{i=0}^{m} c_i x^i \in F[x]$ be a polynomial over F and consider

$$f(T) = \sum_{i=0}^{m} c_i T^i \in \mathcal{L}(V, V)$$

where by convention $T^0 = I$, the identity transformation. If $\alpha \in V_a$, show that

$$\alpha f(T) = f(a)\alpha$$

Conclude that if $f(T) = 0$ and if $\alpha \neq 0$, then a must be a root of the polynomial f.

13.9. Suppose $T^k = 0$ for some integer $k \geq 1$. Find all eigenvalues for T and show that T can be diagonalized if and only if $T = 0$.

13.10. Let $n = \dim_F V$. Prove that there exists a nonzero polynomial $f(x)$ of degree $\leq n^2$ such that $f(T) = 0$. Hint. Recall that $\mathcal{L}(V, V)$ is a vector space over F of dimension n^2 and consider the $n^2 + 1$ linear transformations $T^0 = I, T^1 = T, T^2, \ldots, T^{n^2}$.

CHAPTER III

Determinants

14. Volume Functions

Let V be an n-dimensional vector space over F and let $S\colon V \to V$ be a linear transformation. When is S singular? If $\mathcal{B} = \{\beta_1, \beta_2, \ldots, \beta_n\}$ is a basis for V, then certainly $(\mathcal{B})S = \{\beta_1 S, \beta_2 S, \ldots, \beta_n S\}$ spans the image of S and thus S is singular if and only if the n vectors $\beta_1 S, \beta_2 S, \ldots, \beta_n S$ are linearly dependent. So we change the problem. Given n vectors $\alpha_1, \alpha_2, \ldots, \alpha_n$ in V, when are they linearly dependent? We consider some examples.

EXAMPLE 14.1. Let $V = \mathbb{R}^2$ so that V is the Euclidean plane with each vector corresponding to a point in this plane. Let $\alpha_1, \alpha_2 \in V$ and draw the parallelogram whose vertices include the vectors $0, \alpha_1$ and α_2. The fourth vertex is of course $\alpha_1 + \alpha_2$. Now clearly α_1 and α_2 are

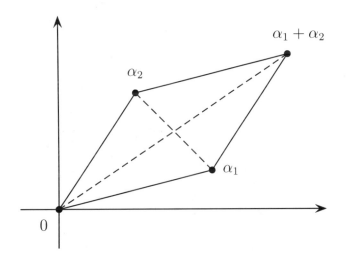

dependent if and only if one of them is an $\mathbb{R}$-multiple of the other and therefore if and only if $0, \alpha_1$ and α_2 are colinear. In this case the parallelogram reduces to a straight line and hence has zero area. Thus the vanishing of the area $v(\alpha_1, \alpha_2)$ of the parallelogram is a necessary and sufficient condition for the linear dependence of α_1 and α_2.

Since the diagonals of a parallelogram bisect each other, we also note that the midpoint of the line segment joining α_1 and α_2 is precisely the point $(1/2)(\alpha_1 + \alpha_2)$, the midpoint of the line joining 0 and $\alpha_1 + \alpha_2$. Of course, this property carries over to $\mathbb{R}^3$ and indeed to all $\mathbb{R}^n$.

EXAMPLE 14.2. Now let $V = \mathbb{R}^3$ so that V is Euclidean 3-space, with each vector again corresponding to a point in this space, and let $\alpha_1, \alpha_2, \alpha_3 \in V$. Here we draw (on the next page) the parallelepiped whose vertices include $0, \alpha_1, \alpha_2$ and α_3. The other vertices are then $\alpha_1 + \alpha_2$, $\alpha_1 + \alpha_3$, $\alpha_2 + \alpha_3$ and $\alpha_1 + \alpha_2 + \alpha_3$. We see easily that $\alpha_1, \alpha_2, \alpha_3$ are linearly dependent if and only if the four points $0, \alpha_1, \alpha_2, \alpha_3$ are coplanar. Indeed, let us suppose for example that α_1 and α_2 are linearly independent, that $\alpha_3 = a_1\alpha_1 + a_2\alpha_2$, and let $\mathcal{P}$ denote the plane

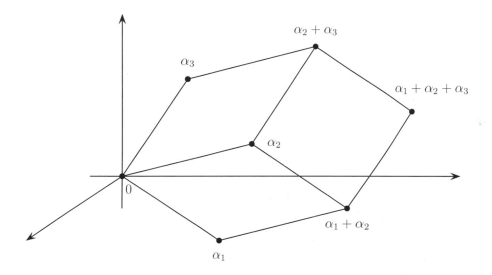

determined by the triangle with vertices 0, α_1 and α_2. For convenience, set $\alpha_1' = 2a_1\alpha_1$ and $\alpha_2' = 2a_2\alpha_2$. Then α_1' is on the line joining 0 to α_1, and α_2' is on the line joining 0 to α_2, so both points are on the plane $\mathcal{P}$. Finally, $\alpha_3 = (1/2)(\alpha_1' + \alpha_2')$ is the midpoint of the line segment joining α_1' and α_2', so α_3 is also on the plane $\mathcal{P}$. Note that this occurs if and only if the volume $v(\alpha_1, \alpha_2, \alpha_3)$ of the parallelepiped is zero.

More generally, if $V = \mathbb{R}^n$ then we are led to consider the n-dimensional volume of the generalized parallelepiped whose vertices include the points $0, \alpha_1, \alpha_2, \ldots, \alpha_n$. Of course, we have no idea as to what the volume really is or in fact whether it exists at all. Indeed, even if we were willing to believe that such a volume exists in $\mathbb{R}^n$, we would still have to come up with something that works for any vector space over any field. So instead, what we will do is use this volume idea as a guide to discover some appropriate function $v(\alpha_1, \alpha_2, \ldots, \alpha_n)$ that will work in general.

Let us consider some properties of n-dimensional volume in $\mathbb{R}^n$. Actually, we will work in $\mathbb{R}^2$ and argue that the analogous situation probably holds in $\mathbb{R}^n$. First, $v(\alpha_1, \alpha_2, \ldots, \alpha_n) \in \mathbb{R}$. Of course we might be tempted to say that the volume is always a nonnegative real number, but this thought quickly dies. One reason is that not all fields have a concept of positive and negative. Additional reasons even for the field $\mathbb{R}$ occur later in this section.

Now how does $v(\alpha_1, \alpha_2, \ldots, \alpha_n)$ behave as a function of each variable individually. Namely, suppose we fix all but the ith vector. We then have a map from V to $\mathbb{R}$ given by $\alpha_i \mapsto v(\alpha_1, \alpha_2, \ldots, \alpha_i, \ldots, \alpha_n)$. What sort of a function is it?

Let $a \in \mathbb{R}$ and consider $v(\alpha_1, \alpha_2, \ldots, a\alpha_i, \ldots, \alpha_n)$. By multiplying α_i by a we have clearly expanded the parallelepiped linearly in one direction by a factor of a and thus we expect the volume to be multiplied by a. In other words

$$v(\alpha_1, \alpha_2, \ldots, a\alpha_i, \ldots, \alpha_n) = a \cdot v(\alpha_1, \alpha_2, \ldots, \alpha_i, \ldots, \alpha_n)$$

This can clearly be seen in the figure below where we have compared $v(3\alpha_1, \alpha_2)$ and $v(\alpha_1, \alpha_2)$. If a is negative, then multiplication by a reverses the sign of a real number. Again negative volumes seem to appear. Of course, we could still multiply $v(\alpha_1, \alpha_2, \ldots, \alpha_i, \ldots, \alpha_n)$ by $|a|$, the absolute value of a, to keep things positive, but the formula is just not as nice and besides, absolute values do not exist in all fields that might be of interest to us.

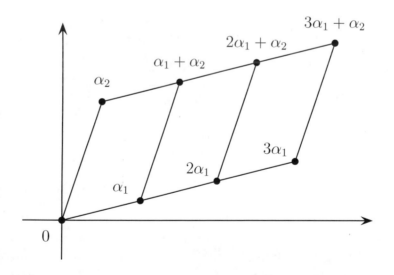

More complicated is the problem of addition of vectors. Namely, what can we say about $v(\alpha_1, \alpha_2, \ldots, \alpha_i + \alpha_i', \ldots, \alpha_n)$? Let us consider $n = 2$ and the area $v(\alpha_1 + \alpha_1', \alpha_2)$ which is the area of the parallelogram $ACC'A'$ in the diagram on the next page. Since triangles ABC and $A'B'C'$ are clearly congruent, we see that $v(\alpha_1 + \alpha_1', \alpha_2)$ is equal to the sum of the areas of the two parallelograms $ABB'A'$ and $BCC'B'$. Of course, the area of $ABB'A'$ is just $v(\alpha_1, \alpha_2)$. Moreover,

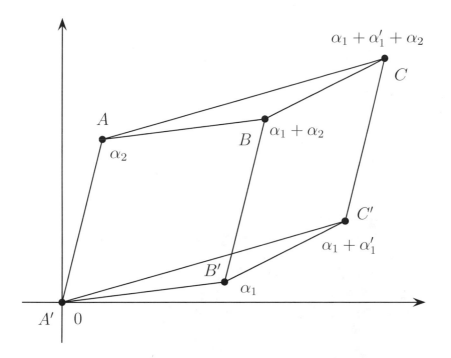

as we see from the diagram below, the area of parallelogram $BCC'B'$ is the same as the area of the congruent parallelogram $DEE'D'$, namely $v(\alpha_1', \alpha_2)$. Thus we have $v(\alpha_1 + \alpha_1', \alpha_2) = v(\alpha_1, \alpha_2) + v(\alpha_1', \alpha_2)$.

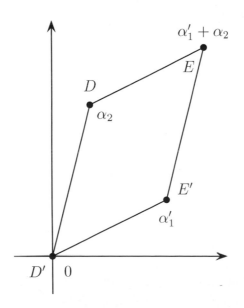

More generally, we would expect that

$$v(\alpha_1, \alpha_2, \ldots, \alpha_i + \alpha_i', \ldots, \alpha_n) = v(\alpha_1, \alpha_2, \ldots, \alpha_i, \ldots, \alpha_n)$$
$$+ v(\alpha_1, \alpha_2, \ldots, \alpha_i', \ldots, \alpha_n)$$

Therefore we see that $v(\alpha_1, \alpha_2, \ldots, \alpha_n)$ viewed as a function strictly of the ith variable is in fact a linear functional from V to $\mathbb{R}$.

Finally, since we have only treated each variable independently so far, we must now consider them together. We know for example that if the vectors $\alpha_1, \alpha_2, \ldots, \alpha_n$ are linearly dependent, then the generalized parallelepiped collapses to an n-dimensional surface rather than a solid and thus has zero volume. For later convenience, we will just record a special case of this. Namely we note that if two vectors α_i and α_j are equal, with $i \neq j$, then $v(\alpha_1, \ldots, \alpha_i, \ldots, \alpha_j, \ldots, \alpha_n) = 0$.

Now we observe that the properties of v listed above translate easily to an arbitrary vector space to give us a *volume function*. Let V be a vector space over F of dimension $n < \infty$. A volume function

$$v \colon \underbrace{V \times V \times \cdots \times V}_{n} \to F$$

is a function of n variable vectors of V with image in F satisfying the following three axioms.

V1. For all $a \in F$ and subscripts i, we have

$$v(\alpha_1, \alpha_2, \ldots, a\alpha_i, \ldots, \alpha_n) = a \cdot v(\alpha_1, \alpha_2, \ldots, \alpha_i, \ldots, \alpha_n)$$

V2. For all i, we have

$$v(\alpha_1, \alpha_2, \ldots, \alpha_i + \alpha_i', \ldots, \alpha_n) = v(\alpha_1, \alpha_2, \ldots, \alpha_i, \ldots, \alpha_n)$$
$$+ v(\alpha_1, \alpha_2, \ldots, \alpha_i', \ldots, \alpha_n)$$

V3. For all $i \neq j$, if $\alpha_i = \alpha_j$ then

$$v(\alpha_1, \ldots, \alpha_i, \ldots, \alpha_j, \ldots, \alpha_n) = 0$$

The first two conditions say that v is *multilinear* while the third condition says that v is *alternating*.

EXAMPLE 14.3. The zero map $v \colon V \times V \times \cdots \times V \to F$ is always trivially a volume function.

EXAMPLE 14.4. If $\dim_F V = n = 1$, then any linear functional $T \colon V \to F$ is a volume function since the third axiom is vacuously satisfied.

EXAMPLE 14.5. Suppose $n = 2$ and let $\{\beta_1, \beta_2\}$ be a fixed basis for V. If $\alpha_1, \alpha_2 \in V$, then $\alpha_1 = a_{11}\beta_1 + a_{12}\beta_2$ and $\alpha_2 = a_{21}\beta_1 + a_{22}\beta_2$ for uniquely determined field elements $a_{ij} \in F$. If we set

$$v(\alpha_1, \alpha_2) = v(a_{11}\beta_1 + a_{12}\beta_2, a_{21}\beta_1 + a_{22}\beta_2) = a_{11}a_{22} - a_{12}a_{21}$$

then it is trivial to see that $v \colon V \times V \to F$ is indeed a volume function.

Let us consider some elementary properties of volume functions. Of course, for all we know at the moment, some vector spaces V may only have the zero volume function. For the remainder of this section, let V be an n-dimensional vector space over the field F and let v be a volume function.

LEMMA 14.1. *With the above notation, we have*

  i. *If $\alpha_i = 0$ for some i, then $v = 0$. In other words,*

$$v(\alpha_1, \ldots, 0, \ldots, \alpha_n) = 0$$

  ii. *If we add a scalar multiple of α_j to α_i, for $j \neq i$, then the value of v does not change. That is,*

$$v(\alpha_1, \ldots, \alpha_i + a\alpha_j, \ldots, \alpha_n) = v(\alpha_1, \ldots, \alpha_i, \ldots, \alpha_n)$$

  iii. *If the vectors α_i and α_j, for $i \neq j$, are interchanged, then the value of v is multiplied by -1. Thus,*

$$v(\alpha_1, \ldots, \alpha_j, \ldots, \alpha_i, \ldots, \alpha_n) = -v(\alpha_1, \ldots, \alpha_i, \ldots, \alpha_j, \ldots, \alpha_n)$$

PROOF. (i) This is just the fact that the image of 0 under a linear transformation is always 0.

(ii) By the axioms for a volume function, we have

$$v(\ldots, \alpha_i + a\alpha_j, \ldots, \alpha_j, \ldots)$$
$$= v(\ldots, \alpha_i, \ldots, \alpha_j, \ldots) + v(\ldots, a\alpha_j, \ldots, \alpha_j, \ldots)$$
$$= v(\ldots, \alpha_i, \ldots, \alpha_j, \ldots) + a{\cdot}v(\ldots, \alpha_j, \ldots, \alpha_j, \ldots)$$
$$= v(\ldots, \alpha_i, \ldots, \alpha_j, \ldots)$$

(iii) Let us replace the ith and jth variables by $\alpha_i + \alpha_j$. Then by linearity and (ii) above, we have

$$0 = v(\ldots, \alpha_i + \alpha_j, \ldots, \alpha_i + \alpha_j, \ldots)$$
$$= v(\ldots, \alpha_i + \alpha_j, \ldots, \alpha_i, \ldots) + v(\ldots, \alpha_i + \alpha_j, \ldots, \alpha_j, \ldots)$$
$$= v(\ldots, \alpha_j, \ldots, \alpha_i, \ldots) + v(\ldots, \alpha_i, \ldots, \alpha_j, \ldots)$$

and the result follows.   □

LEMMA 14.2. *If the vectors $\alpha_1, \alpha_2, \ldots, \alpha_n$ are linearly dependent, then $v(\alpha_1, \alpha_2, \ldots, \alpha_n) = 0$.*

PROOF. Since the vectors are linearly dependent, for some i, we can solve for α_i in terms of the remaining α_j and obtain $\alpha_i + \sum_{j \neq i} a_j \alpha_j = 0$. We now apply part (ii) of the previous lemma and successively add $a_1 \alpha_1, a_2 \alpha_2, \dots, a_n \alpha_n$ (for $j \neq i$) to the ith variable without changing v. Thus

$$v(\alpha_1, \dots, \alpha_i, \dots, \alpha_n) = v(\alpha_1, \dots, \alpha_i + \sum_{j \neq i} a_j \alpha_j, \dots, \alpha_n)$$

$$= v(\alpha_1, \dots, 0, \dots, \alpha_n) = 0$$

and the lemma is proved.   □

This solves part of the problem. Namely if the n vectors are linearly dependent then the volume function vanishes. What happens if the vectors are linearly independent? We will show below that if v is not the identically zero function, then v does not vanish on any linearly independent set.

LEMMA 14.3. *Suppose σ is any permutation of the set $\{1, 2, \dots, n\}$. Then*

$$v(\alpha_{\sigma(1)}, \dots, \alpha_{\sigma(i)}, \dots, \alpha_{\sigma(n)}) = \pm v(\alpha_1, \dots, \alpha_i, \dots, \alpha_n)$$

where the $\pm$ sign is determined solely by σ.

PROOF. By Lemma 14.1(iii), we can interchange any two entries and only change v by a factor of -1. So we first interchange α_1 with $\alpha_{\sigma(1)}$ if $1 \neq \sigma(1)$ so that $v(\alpha_{\sigma(1)}, \dots, \alpha_{\sigma(i)}, \dots, \alpha_{\sigma(n)})$ now has α_1 as its first variable. Then we interchange α_2 if necessary with the second entry and continue this process. At each step we either leave v alone or multiply it by -1. This clearly yields the result since the $\pm$ sign is determined in this way by σ.   □

It is important to observe that the choice of the $\pm$ sign above depends on our specific procedure for reordering the α_is. There is certainly no a priori reason to believe that a different reordering scheme will yield the same sign. Indeed this is the fundamental problem in trying to define a nontrivial volume function. The following argument may look complicated, but it really a simple application of the multilinearity of v.

THEOREM 14.1. *Let V be a vector space over the field F having dimension $n < \infty$. Let $\mathcal{A} = \{\alpha_1, \alpha_2, \dots, \alpha_n\}$ be a subset of V and let $\mathcal{B} = \{\beta_1, \beta_2, \dots, \beta_n\}$ be a basis. If v is a volume function then there exists $c \in F$ depending only upon $\mathcal{A}$ and $\mathcal{B}$ with*

$$v(\alpha_1, \alpha_2, \dots, \alpha_n) = c \cdot v(\beta_1, \beta_2, \dots, \beta_n)$$

PROOF. Since $\mathcal{B}$ is a basis for V, we can write

$$\alpha_1 = a_{11}\beta_1 + a_{12}\beta_2 + \cdots + a_{1n}\beta_n = \sum_j a_{1j}\beta_j$$

$$\alpha_2 = a_{21}\beta_1 + a_{22}\beta_2 + \cdots + a_{2n}\beta_n = \sum_j a_{2j}\beta_j$$

$$\cdots \cdots \cdots \cdots$$

$$\alpha_n = a_{n1}\beta_1 + a_{n2}\beta_2 + \cdots + a_{nn}\beta_n = \sum_j a_{nj}\beta_j$$

for uniquely determined $a_{ij} \in F$. Then

$$v(\alpha_1, \alpha_2, \ldots, \alpha_n) = v\left(\sum_j a_{1j}\beta_j, \sum_j a_{2j}\beta_j, \ldots, \sum_j a_{nj}\beta_j\right)$$

Now v is a linear transformation when viewed as a function of each variable independently so that

$$v\left(\ldots, \sum_j a_{ij}\beta_j, \ldots\right) = \sum_j a_{ij} \cdot v(\ldots, \beta_j, \ldots)$$

Thus by expanding $v(\alpha_1, \alpha_2, \ldots, \alpha_n)$ in this manner, we get a sum of n terms from each variable and we therefore obtain

$$v(\alpha_1, \alpha_2, \ldots, \alpha_n) = \sum_{\substack{\text{all } n-\text{tuples} \\ (j_1, j_2, \ldots, j_n)}} a_{1j_1} a_{2j_2} \cdots a_{nj_n} \cdot v(\beta_{j_1}, \beta_{j_2}, \ldots, \beta_{j_n})$$

a sum of n^n terms.

Consider one such term

$$a_{1j_1} a_{2j_2} \cdots a_{nj_n} \cdot v(\beta_{j_1}, \beta_{j_2}, \ldots, \beta_{j_n})$$

Since there are precisely n vectors β_j and precisely n entries, we see that either $j_1, j_2, \ldots, j_n$ is a permutation of the numbers $1, 2, \ldots, n$ or else two of the j_is are equal. Of course, in the latter case we see that $v(\beta_{j_1}, \beta_{j_2}, \ldots, \beta_{j_n}) = 0$ and the entire term vanishes. Thus, by deleting these zero terms, we have

$$v(\alpha_1, \alpha_2, \ldots, \alpha_n) = \sum_\sigma a_{1\sigma(1)} a_{2\sigma(2)} \cdots a_{n\sigma(n)} \cdot v(\beta_{\sigma(1)}, \beta_{\sigma(2)}, \ldots, \beta_{\sigma(n)})$$

where the sum is over all $n!$ permutations σ of the set $\{1, 2, \ldots, n\}$. Finally, by the previous lemma

$$v(\beta_{\sigma(1)}, \beta_{\sigma(2)}, \ldots, \beta_{\sigma(n)}) = \pm v(\beta_1, \beta_2, \ldots, \beta_n)$$

where the $\pm$ sign is determined by σ. Thus we get

$$v(\alpha_1, \alpha_2, \ldots, \alpha_n) = \left(\sum_\sigma \pm a_{1\sigma(1)} a_{2\sigma(2)} \cdots a_{n\sigma(n)} \right) \cdot v(\beta_1, \beta_2, \ldots, \beta_n)$$

$$= c \cdot v(\beta_1, \beta_2, \ldots, \beta_n)$$

where

$$c = \sum_\sigma \pm a_{1\sigma(1)} a_{2\sigma(2)} \cdots a_{n\sigma(n)}$$

depends only on $\mathcal{A}$ and $\mathcal{B}$. The theorem is proved. $\square$

As an immediate corollary we have

COROLLARY 14.1. *Let $\mathcal{B} = \{\beta_1, \beta_2, \ldots, \beta_n\}$ be a basis for V and let v be a volume function. Then v is uniquely determined by the value $v(\beta_1, \beta_2, \ldots, \beta_n)$. In particular if $v(\beta_1, \beta_2, \ldots, \beta_n) = 0$, then v is identically 0.*

Problems

14.1. Verify that the function defined in Example 14.5 is a volume function.

Let V be a 3-dimensional vector space over F with basis $\mathcal{B} = \{\beta_1, \beta_2, \beta_3\}$. Let $\alpha_1, \alpha_2, \alpha_3 \in V$ and let v be a volume function.

14.2. Each of the following is equal to $\pm v(\alpha_1, \alpha_2, \alpha_3)$.

$$v(\alpha_1, \alpha_2, \alpha_3), \qquad v(\alpha_2, \alpha_3, \alpha_1), \qquad v(\alpha_3, \alpha_1, \alpha_2)$$
$$v(\alpha_1, \alpha_3, \alpha_2), \qquad v(\alpha_2, \alpha_1, \alpha_3), \qquad v(\alpha_3, \alpha_2, \alpha_1)$$

Find the correct sign for each.

14.3. Follow the proof of Theorem 14.1 in this special case and compute c explicitly.

14.4. Let v' be the function given by $v'(\alpha_1, \alpha_2, \alpha_3) = c$ where c is given as in the previous exercise. Prove that v' is a volume function with $v'(\beta_1, \beta_2, \beta_3) = 1$.

Now let V be an n-dimensional vector space over $\mathbb{R}$ and let v be a volume function. Suppose σ is a permutation of the set $\{1, 2, \ldots, n\}$.

14.5. If $\alpha_{\sigma(1)}, \alpha_{\sigma(2)}, \ldots, \alpha_{\sigma(n)}$ can be reordered to $\alpha_1, \alpha_2, \ldots, \alpha_n$ in t interchanges, show that

$$v(\alpha_{\sigma(1)}, \alpha_{\sigma(2)}, \ldots, \alpha_{\sigma(n)}) = (-1)^t \cdot v(\alpha_1, \alpha_2, \ldots, \alpha_n)$$

14.6. Suppose v exists with $v(\alpha_1, \alpha_2, \ldots, \alpha_n) \neq 0$ for some vectors $\alpha_1, \alpha_2, \ldots, \alpha_n \in V$. Show that the $\pm$ sign in

$$v(\alpha_{\sigma(1)}, \alpha_{\sigma(2)}, \ldots, \alpha_{\sigma(n)}) = \pm v(\alpha_1, \alpha_2, \ldots, \alpha_n)$$

is uniquely determined. What can you conclude about the parity, that is the oddness or evenness, of the number of interchanges required in the reordering process.

14.7. Suppose $n = 5$. Find three different reordering processes for each of $v(\alpha_2, \alpha_3, \alpha_4, \alpha_5, \alpha_1)$ and $v(\alpha_3, \alpha_1, \alpha_2, \alpha_5, \alpha_4)$. Count the number of interchanges in each case and see whether the parity remains the same.

14.8. Let V be a vector space over F with basis $\{\beta_1, \beta_2, \ldots, \beta_n\}$ and let v' be a nonzero volume function. Show that, for each $a \in F$, V has one and only one volume function v with $v(\beta_1, \beta_2, \ldots, \beta_n) = a$.

14.9. Suppose $\mathbb{R}^n$ has an n-dimensional volume v. Explain why v is unique up to a scalar factor.

14.10. Find the volumes of the geometric figures described in Examples 14.1 and 14.2.

15. Determinants

We seem to know everything about volume functions except whether nonzero ones exist. In this section, we prove their existence. There are several possible ways of doing this. One approach is to study the $\pm$ sign that occurs in

$$v(\alpha_{\sigma(1)}, \alpha_{\sigma(2)}, \ldots, \alpha_{\sigma(n)}) = \pm v(\alpha_1, \alpha_2, \ldots, \alpha_n)$$

Indeed, once this sign is properly understood, we would have no difficulty in defining v. A second approach and the one we take here is inductive.

Let α be an $n \times n$ matrix over F. Then each row of α is an n-tuple of elements of F and hence can be thought of as being a vector in F^n. In this way, α is composed on n vectors of F^n and thus any function sending α to F can be thought of as a function of n elements of F^n. We can therefore translate the notion of a volume function to this situation and we call the resulting function the *determinant*.

Formally the determinant is a map $\det\colon F^{n\times n} \to F$ satisfying the following three axioms.

D1. If we consider $\det \alpha$ as a function of the ith row of α, for any i, with all other rows kept fixed, then det is a linear transformation from F^n to F.

D2. If two rows of α are identical, then $\det \alpha = 0$.

D3. $\det I_n = 1$.

The first two above tell us, as we have already indicated, that det is a volume function on F^n. The third axiom is a normalization. It says that the volume function evaluated at the basis $\{\beta_1, \beta_2, \ldots, \beta_n\}$, where $\beta_i = (0, 0, \ldots, 1, \ldots, n)$ has a 1 in the ith spot, is equal to 1. Thus, by Corollary 14.1 we have

THEOREM 15.1. *If* $\det\colon F^{n\times n} \to F$ *exists, then it is unique.*

We now proceed to show that det does in fact exist by induction on n. If $\alpha \in F^{n\times n}$ then we denote by α_{ij} the submatrix of α obtained by deleting the ith row and jth column. Clearly $\alpha_{ij} \in F^{(n-1)\times(n-1)}$.

LEMMA 15.1. *Let* $n > 1$ *and suppose that the determinant map* $\det\colon F^{(n-1)\times(n-1)} \to F$ *exists. Fix an integer* s *with* $1 \le s \le n$ *and define a function* $d_s\colon F^{n\times n} \to F$ *by*

$$d_s(\alpha) = \sum_{i=1}^{n} (-1)^{i+s} a_{is} \det \alpha_{is}$$

where $\alpha = [a_{ij}]$. Then $d_s(\alpha)$ is an $n \times n$ determinant map.

PROOF. We obviously have to check that d_s satisfies all three axioms above. This is not difficult, just tedious. We consider each axiom in turn.

D1) Fix a row subscript k. We study d_s as a function of the kth row of α. Thus here all matrices of interest will have the same fixed entries in all rows but the kth. We can indicate this symbolically by

$$\alpha = \begin{bmatrix} e_{ij} \\ a_{kj} \\ e_{ij} \end{bmatrix}$$

In other words, all the ij-entries for $i \neq k$ are e_{ij} and these are fixed. The kj-entries are a_{kj} and we will allow these to vary.

Let $b \in F$ and set

$$\alpha = \begin{bmatrix} e_{ij} \\ a_{kj} \\ e_{ij} \end{bmatrix}, \qquad \beta = \begin{bmatrix} e_{ij} \\ b \cdot a_{kj} \\ e_{ij} \end{bmatrix}$$

Then by definition

$$d_s(\beta) = (-1)^{k+s}(b \cdot a_{ks}) \det \beta_{ks} + \sum_{i \neq k} (-1)^{i+s} e_{is} \det \beta_{is}$$

Now for $i \neq k$, the kth row is not deleted in β_{is}. Thus α_{is} and β_{is} agree in all rows but one and in that row all entries of β_{is} are equal to b times the corresponding entry in α_{is}. By properties of det we therefore have $\det \beta_{is} = b(\det \alpha_{is})$ for $i \neq k$. On the other hand, for $i = k$, the kth row is deleted so $\beta_{ks} = \alpha_{ks}$. This yields

$$d_s(\beta) = (-1)^{k+s}(b \cdot a_{ks}) \det \alpha_{ks} + \sum_{i \neq k} (-1)^{i+s} e_{is} \cdot b(\det \alpha_{is})$$

$$= b\Big((-1)^{k+s} a_{ks} \det \alpha_{ks} + \sum_{i \neq k} (-1)^{i+s} e_{is} \det \alpha_{is} \Big)$$

$$= b \cdot d_s(\alpha)$$

Now let α, β and γ be given by

$$\alpha = \begin{bmatrix} e_{ij} \\ a_{kj} \\ e_{ij} \end{bmatrix}, \qquad \beta = \begin{bmatrix} e_{ij} \\ b_{kj} \\ e_{ij} \end{bmatrix}, \qquad \gamma = \begin{bmatrix} e_{ij} \\ a_{kj} + b_{kj} \\ e_{ij} \end{bmatrix}$$

Then

$$d_s(\gamma) = (-1)^{k+s}(a_{ks} + b_{ks}) \det \gamma_{ks} + \sum_{i \neq k} (-1)^{i+s} e_{is} \det \gamma_{is}$$

Now for $i \neq k$, the kth row is not deleted in γ_{is}. Clearly α_{is}, β_{is} and γ_{is} agree in all the other rows and they add appropriately in this row. Thus

$$\det \gamma_{is} = \det \alpha_{is} + \det \beta_{is}$$

for $i \neq k$. On the other hand, for $i = k$ we have $\gamma_{ks} = \alpha_{ks} = \beta_{ks}$. This yields

$$d_s(\gamma) = (-1)^{k+s} a_{ks} \det \alpha_{ks} + \sum_{i \neq k} (-1)^{i+s} e_{is} \det \alpha_{is}$$

$$+ (-1)^{k+s} b_{ks} \det \beta_{ks} + \sum_{i \neq k} (-1)^{i+s} e_{is} \det \beta_{is}$$

$$= d_s(\alpha) + d_s(\beta)$$

and d_s satisfies the first axiom.

D2) Let us suppose that the kth and ℓth rows of α are identical and say $\ell > k$. If $\alpha = [a_{ij}]$, then

$$d_s(\alpha) = \sum_i (-1)^{i+s} a_{is} \det \alpha_{is}$$

Now if $i \neq k$ or ℓ, then neither row is deleted in α_{is}. This means that α_{is} has two identical rows and $\det \alpha_{is} = 0$. Therefore only the k and ℓ terms occur in the above sum and since $a_{ks} = a_{\ell s}$, we have

$$d_s(\alpha) = (-1)^{k+s} a_{ks} \left(\det \alpha_{ks} + (-1)^{\ell-k} \det \alpha_{\ell s} \right)$$

How do α_{ks} and $\alpha_{\ell s}$ compare? In one case we deleted the kth row and in the other the ℓth. But since these rows are identical, we see that α_{ks} and $\alpha_{\ell s}$ have the same rows occurring but in a slightly different order. In fact, we have pictorially

$$\alpha_{ks} = \begin{bmatrix} \text{row } 1 \\ \text{row } 2 \\ \vdots \\ \text{row } k \text{ (missing)} \\ \vdots \\ \text{row } \ell \\ \vdots \\ \text{row } n \end{bmatrix} \qquad \alpha_{\ell s} = \begin{bmatrix} \text{row } 1 \\ \text{row } 2 \\ \vdots \\ \text{row } k \\ \vdots \\ \text{row } \ell \text{ (missing)} \\ \vdots \\ \text{row } n \end{bmatrix}$$

Since det is a volume function on F^{n-1}, interchanging rows merely multiplies the value det by -1. Now we make $\alpha_{\ell s}$ identical to α_{ks} by slowly lowering row k to the ℓ position. We do this by consecutively interchanging the kth row with rows $k+1, k+2, \ldots, \ell-1$.

$$\alpha_{\ell s} = \begin{bmatrix} \vdots \\ \text{row } k \\ \text{row } k+1 \\ \text{row } k+2 \\ \vdots \\ \text{row } \ell-2 \\ \text{row } \ell-1 \\ \text{row } \ell \text{ (missing)} \\ \vdots \end{bmatrix}$$

By doing it this way, we have not changed the ordering of the other rows and yet we have shifted row k down to position ℓ. Moreover this was achieved in $\ell - k - 1$ interchanges (since, for example if $k = \ell - 1$ then no interchanges are needed). Thus we have

$$\det \alpha_{\ell s} = (-1)^{\ell-k-1} \det \alpha_{ks}$$

and

$$d_s(\alpha) = (-1)^{k+s} a_{ks} \Big(\det \alpha_{ks} + (-1)^{\ell-k}(-1)^{\ell-k-1} \det \alpha_{ks} \Big)$$
$$= (-1)^{k+s} a_{ks} \Big(\det \alpha_{ks} - \det \alpha_{ks} \Big) = 0$$

as required.

D3) If $I_n = \alpha = [a_{ij}]$ then

$$d_s(I_n) = \sum_i (-1)^{i+s} a_{is} \det \alpha_{is}$$

Now $a_{is} = 0$ for $i \neq s$ and $a_{ss} = 1$, so

$$d_s(I_n) = (-1)^{s+s} a_{ss} \det \alpha_{ss} = \det \alpha_{ss}$$

Since clearly $\alpha_{ss} = I_{n-1}$ and $\det I_{n-1} = 1$, this axiom is satisfied. We have therefore shown that the function d_s satisfies all the appropriate axioms and thus d_s is a determinant function on $F^{n \times n}$. □

It is now an easy matter to prove that determinants exist.

THEOREM 15.2. *The determinant map* $\det \colon F^{n \times n} \to F$ *exists for all integers n. Moreover for $n > 1$ and for any integer s with $1 \leq s \leq n$, we have*

$$\det \alpha = \sum_{i=1}^{n} (-1)^{i+s} a_{is} \det \alpha_{is}$$

where $\alpha = [a_{ij}] \in F^{n \times n}$. The latter is called the cofactor expansion of the determinant with respect to the sth column.

PROOF. We proceed by induction on n. If $n = 1$ then $\det[a] = a$ is easily seen to satisfy all the appropriate axioms.

Now suppose $\det\colon F^{(n-1) \times (n-1)} \to F$ exists. Fix s with $1 \leq s \leq n$ and define $d_s\colon F^{n \times n} \to F$ by

$$d_s(\alpha) = \sum_{i=1}^{n} (-1)^{i+s} a_{is} \det \alpha_{is}$$

where $\alpha = [a_{ij}]$. By the preceding lemma d_s is a determinant function and hence by the uniqueness Theorem 15.1, d_s is the determinant function. Thus we have shown that $\det\colon F^{n \times n} \to F$ exists and that $\det = d_s$ for any suitable integer s. The theorem now follows by induction. $\square$

COROLLARY 15.1. *Let V be an n-dimensional vector space over F with basis $\{\beta_1, \beta_2, \ldots, \beta_n\}$. Define*

$$v\colon \underbrace{V \times V \times \cdots \times V}_{n} \to F$$

by $v(\alpha_1, \alpha_2, \ldots, \alpha_n) = \det[a_{ij}]$ where $\alpha_i = \sum_j a_{ij}\beta_j$. Then v is a nonzero volume function with $v(\beta_1, \beta_2, \ldots, \beta_n) = 1$.

PROOF. We must show that v satisfies all the axioms for a volume function. Suppose first that α_k is replaced by $b\alpha_k$. Then clearly the kth row of $[a_{ij}]$ is multiplied by $b \in F$ so (D1) yields

$$v(\alpha_1, \ldots, b\alpha_k, \ldots, \alpha_n) = b \det[a_{ij}] = b{\cdot}v(\alpha_1, \ldots, \alpha_k, \ldots, \alpha_n)$$

Secondly, consider $v(\alpha_1, \ldots, \alpha_k + \alpha_k', \ldots, \alpha_n)$ where $\alpha_k = \sum_j a_{kj}\beta_j$ and $\alpha_k' = \sum_j a_{kj}'\beta_j$. Then $\alpha_k + \alpha_k' = \sum_j (a_{kj} + a_{kj}')\beta_j$. This means that the corresponding matrices for the definition of $v(\alpha_1, \ldots, \alpha_k, \ldots, \alpha_n)$ and $v(\alpha_1, \ldots, \alpha_k', \ldots, \alpha_n)$ agree except in the kth row and there they add to give the kth row of the matrix for $v(\alpha_1, \ldots, \alpha_k + \alpha_k', \ldots, \alpha_n)$. Thus again by (D1) we have

$$v(\alpha_1, \ldots, \alpha_k + \alpha_k', \ldots, \alpha_n) =$$
$$v(\alpha_1, \ldots, \alpha_k, \ldots, \alpha_n) + v(\alpha_1, \ldots, \alpha_k', \ldots, \alpha_n)$$

Next, if the two vectors α_r and α_s are identical, then certainly the rth and sth rows of $[a_{ij}]$ are identical so

$$v(\alpha_1, \ldots, \alpha_r, \ldots, \alpha_s, \ldots, \alpha_n) = \det[a_{ij}] = 0$$

We have therefore shown that v is a volume function.

Finally

$$v(\beta_1, \beta_2, \ldots, \beta_n) = \det I_n = 1$$

so v is nonzero and the result follows.   □

COROLLARY 15.2. *Let $S\colon V \to V$ be a linear transformation and let $\mathcal{B}$ and $\mathcal{B}'$ be bases for V. Then S is singular if and only if $\det {}_{\mathcal{B}'}[S]_{\mathcal{B}} = 0$.*

PROOF. Let the volume function v be defined as in the preceding corollary using the basis $\mathcal{B} = \{\beta_1, \beta_2, \ldots, \beta_n\}$. Let $\mathcal{B}' = \{\beta_1', \beta_2', \ldots, \beta_n'\}$ and set $\alpha_i = \beta_i'S$, the image of β_i' under S. If

$$\beta_i'S = \alpha_i = \sum_j a_{ij}\beta_j$$

then ${}_{\mathcal{B}'}[S]_{\mathcal{B}} = [a_{ij}]$ and

$$\det {}_{\mathcal{B}'}[S]_{\mathcal{B}} = \det[a_{ij}] = v(\alpha_1, \alpha_2, \ldots, \alpha_n)$$

Now $(\mathcal{B}')S = \{\alpha_1, \alpha_2, \ldots, \alpha_n\}$ spans $\operatorname{im} S$. Thus S is singular if and only if $\{\alpha_1, \alpha_2, \ldots, \alpha_n\}$ is linearly dependent and hence if and only if $v(\alpha_1, \alpha_2, \ldots, \alpha_n) = 0$ since v is a nonzero volume function. This completes the proof.   □

If $\alpha = [a_{ij}]$ is an $n \times n$ matrix, then we will use the notation

$$\det \alpha = \det[a_{ij}] = |a_{ij}| = \begin{vmatrix} a_{11} & a_{12} & \cdots & a_{1n} \\ a_{21} & a_{22} & \cdots & a_{2n} \\ & & \cdots \cdots \cdots & \\ a_{n1} & a_{n2} & \cdots & a_{nn} \end{vmatrix}$$

Now the proof of the existence of the determinant is actually constructive. It really gives us a way of computing det for at least small values of n. Note that the vertical lines here stand for determinant and not absolute value.

EXAMPLE 15.1. If $n = 1$, then $|a_{11}| = a_{11}$. Let $n = 2$. Then using the cofactor expansion with respect to the first column, we have

$$\begin{vmatrix} a_{11} & a_{12} \\ a_{21} & a_{22} \end{vmatrix} = a_{11}|a_{22}| - a_{21}|a_{12}|$$

$$= a_{11}a_{22} - a_{12}a_{21}$$

a not unexpected result.

Suppose $n = 3$ and again use the cofactor expansion down the first column. Then

$$\begin{vmatrix} a_{11} & a_{12} & a_{13} \\ a_{21} & a_{22} & a_{23} \\ a_{31} & a_{32} & a_{33} \end{vmatrix} = a_{11} \begin{vmatrix} a_{22} & a_{23} \\ a_{32} & a_{33} \end{vmatrix} - a_{21} \begin{vmatrix} a_{12} & a_{13} \\ a_{32} & a_{33} \end{vmatrix}$$

$$+ a_{31} \begin{vmatrix} a_{12} & a_{13} \\ a_{22} & a_{23} \end{vmatrix}$$

$$= a_{11}(a_{22}a_{33} - a_{32}a_{23}) - a_{21}(a_{12}a_{33} - a_{32}a_{13})$$

$$+ a_{31}(a_{12}a_{23} - a_{22}a_{13})$$

$$= a_{11}a_{22}a_{33} - a_{11}a_{23}a_{32} + a_{13}a_{21}a_{32}$$

$$- a_{12}a_{21}a_{33} + a_{12}a_{23}a_{31} - a_{13}a_{22}a_{31}$$

Finally, we have

THEOREM 15.3. *Let* $[a_{ij}] \in F^{n \times n}$. *Then*

$$\det[a_{ij}] = |a_{ij}| = \sum \pm a_{1j_1} a_{2j_2} \cdots a_{nj_n}$$

where the sum is over all n-tuples $(j_1, j_2, \ldots, j_n)$ *of distinct column subscripts. Here the* $\pm$ *sign is uniquely determined and the main diagonal term* $a_{11}a_{22} \cdots a_{nn}$ *occurs with a plus sign.*

PROOF. We proceed by induction on n. We have already seen in the above example that the result is true for $n = 1, 2, 3$. Suppose that the result holds for $(n-1) \times (n-1)$ matrices and let $\alpha = [a_{ij}] \in F^{n \times n}$. Then by induction

$$\det \alpha_{k1} = \sum \pm a_{1j_1} \cdots a_{(k-1),j_{(k-1)}} a_{(k+1),j_{(k+1)}} \cdots a_{nj_n}$$

where the sum is over all $(n-1)$-tuples $(j_1, \ldots, j_{(k-1)}, j_{(k+1)}, \ldots, j_n)$ of distinct integers between 2 and n. This follows because the row subscripts of α_{k1} omit k and the column subscripts omit 1. Applying the cofactor expansion of $\det \alpha$ with respect to the first column, we therefore get

$$\det \alpha = \sum_k (-1)^{k+1} a_{k1} \cdot \det \alpha_{k1}$$

$$= \sum_k (-1)^{k+1} a_{k1} \sum \pm a_{1j_1} \cdots a_{(k-1),j_{(k-1)}} a_{(k+1),j_{(k+1)}} \cdots a_{nj_n}$$

$$= \sum \pm a_{1j_1} a_{2j_2} \cdots a_{kj_k} \cdots a_{nj_n}$$

Finally the term $a_{11}a_{22} \cdots a_{nn}$ clearly comes from $(-1)^{1+1} a_{11}$ times the diagonal term $a_{22}a_{33} \cdots a_{nn}$ of $\det \alpha_{11}$ and hence it occurs with a plus sign. The result follows. $\qquad \square$

Problems

15.1. Evaluate the determinant of $\alpha \in \mathbb{R}^{4 \times 4}$ where

$$\alpha = \begin{bmatrix} 2 & 1 & 0 & -1 \\ -1 & 0 & 1 & 3 \\ 0 & 2 & 7 & 1 \\ 0 & 0 & -1 & 2 \end{bmatrix}$$

15.2. Let $\alpha \in \mathbb{R}^{3 \times 3}$ be given by

$$\alpha = \begin{bmatrix} 3 & -1 & 2 \\ 0 & 2 & 4 \\ -1 & 0 & 1 \end{bmatrix}$$

Evaluate the determinants of α and of α^T, the transpose of α. Compare your results.

15.3. Let $\alpha, \beta \in \mathbb{R}^{2 \times 2}$ be given by

$$\alpha = \begin{bmatrix} 4 & 1 \\ 1 & 2 \end{bmatrix} \qquad \beta = \begin{bmatrix} 2 & -1 \\ 1 & 1 \end{bmatrix}$$

Evaluate $\det \alpha$, $\det \beta$, $\det(\alpha\beta)$ and $\det(\beta\alpha)$, Compare your results.

15.4. Recall Example 13.6 where $S \colon V \to V$ is given by

$$_A[S]_A - \begin{bmatrix} a+10 & -9 & -6 \\ -32 & a+23 & 18 \\ 61 & -47 & a-35 \end{bmatrix}$$

Find $\det {}_A[S]_A$. For what values of $a \in \mathbb{R}$ is S singular?

15.5. Let $\alpha \in F^{5 \times 5}$ be given by

$$\alpha = \begin{bmatrix} a_{11} & a_{12} & a_{13} & a_{14} & a_{15} \\ 0 & a_{22} & a_{23} & a_{24} & a_{25} \\ 0 & 0 & a_{33} & a_{34} & a_{35} \\ 0 & 0 & 0 & a_{44} & a_{45} \\ 0 & 0 & 0 & 0 & a_{55} \end{bmatrix}$$

Find $\det \alpha$.

15.6. Let $\beta \in F^{n \times n}$. Prove that β is nonsingular if and only if $\det \beta \neq 0$.

15.7. Let $\alpha, \beta \in F^{n \times n}$. Is the formula

$$\det(\alpha + \beta) = \det \alpha + \det \beta$$

true in general?

Let V be a vector space over F of dimension n and let v be a nonzero volume function. Let $T\colon V \to V$ be a linear transformation.

15.8. Define a function
$$vT\colon \underbrace{V \times V \times \cdots \times V}_{n} \to F$$
by
$$(vT)(\alpha_1, \alpha_2, \ldots, \alpha_n) = v(\alpha_1 T, \alpha_2 T, \ldots, \alpha_n T)$$
Show that vT is a volume function.

15.9. If $a \in F$ define av by
$$(av)(\alpha_1, \alpha_2, \ldots, \alpha_n) = a \cdot (v(\alpha_1, \alpha_2, \ldots, \alpha_n))$$
Show that av is a volume function and that every volume function on V is of this form for a unique $a \in F$.

15.10. From the above we see that there exists a unique $c \in F$ with $vT = cv$. Prove that $c = 0$ if and only if T is singular.

16. Consequences of Uniqueness

At this point we could use the existence of determinants to further our study of eigenvalues. However, instead we will devote this section to a consideration of certain additional properties of determinants. As we will see, the theorems all follow beautifully from the uniqueness aspect which we repeat below.

LEMMA 16.1. *Let* $d\colon F^{n\times n} \to F$ *satisfy*

    i. d is a linear transformation from F^n to F when viewed as a function of any particular row of the matrix.
    ii. $d(\alpha) = 0$ if two rows of α are identical.

Then for all α, we have

$$d(\alpha) = c\cdot(\det \alpha)$$

where $c = d(I_n)$.

PROOF. Observe that the function $d'\colon F^{n\times n} \to F$ given by $d'(\alpha) = d(I_n)\cdot \det \alpha$ satisfies (i) and (ii), and $d(I_n) = d'(I_n)$. Thus d and d' are volume functions on F^n that agree on a basis. By the uniqueness theorem, Corollary 14.1, d and d' are identical. $\qquad\square$

Now we have treated $\det \alpha$ as a function of the rows of α. How does it behave as a column function? The answer is that it behaves in just the same way. Observe that the columns are merely n-tuples on their side, so we can also consider them as elements of F^n.

LEMMA 16.2. *Let* $\det\colon F^{n\times n} \to F$. *Then* $\det$ *is a linear transformation from F^n to F when viewed strictly as a function of one particular column. Moreover, if two columns of α are identical, then $\det \alpha = 0$.*

PROOF. This all follows from the expansion of det by cofactors down a particular column. Let us consider $\det \alpha$ as a function of the kth column. If $\alpha = [a_{ij}]$ then

$$\det \alpha = \sum_{i=1}^{n}(-1)^{i+k}a_{ik}\det \alpha_{ik}$$

Observe that if we fix all the columns of α other than the kth, then α_{ik} is kept fixed and the above formula is just

$$\det \alpha = \sum_{i=1}^{n}a_{ik}c_i$$

for some constants $c_i \in F$. Since the map $F^n \to F$ given by

$$(a_{1k}, a_{2k}, \ldots, a_{nk}) \mapsto \sum_{i=1}^{n} a_{ik} c_i$$

is clearly a linear functional, the first part of the lemma is proved.

We prove the second part by induction on n. The result is vacuously true for $n = 1$. Now let $n = 2$. Then

$$\begin{vmatrix} a_{11} & a_{12} \\ a_{21} & a_{22} \end{vmatrix} = a_{11}a_{22} - a_{12}a_{21}$$

so if the two columns are identical, then $a_{11} = a_{12}$, $a_{21} = a_{22}$ and $|a_{ij}| = 0$. Finally, let $n > 2$ and suppose the result is true for $n - 1$. Let $\alpha \in F^{n \times n}$ with say its rth and sth columns identical. Since $n \geq 3$, we can choose $k \neq r, s$ and then by expanding $\det \alpha$ with respect to the kth column we have

$$\det \alpha = \sum_{i=1}^{n} (-1)^{i+k} a_{ik} \det \alpha_{ik}$$

where $\alpha = [a_{ij}]$. Now in obtaining α_{ik}, we deleted neither the rth nor the sth column. Thus α_{ik} has two columns identical and by induction $\det \alpha_{ik} = 0$ for all i. Hence $\det \alpha = 0$ and the lemma is proved. $\square$

Let $\alpha = [a_{ij}] \in F^{n \times n}$. Recall that the transpose α^{T} of α is defined to be the $n \times n$ matrix $\alpha^{\mathsf{T}} = [a'_{ij}]$ where $a'_{ij} = a_{ji}$. In other words, α^{T} is obtained from α by reflecting it about the main diagonal. In this process we see that rows and columns are interchanged. But one thing that is not changed is the determinant.

THEOREM 16.1. *Let $\alpha \in F^{n \times n}$. Then*

$$\det \alpha^{\mathsf{T}} = \det \alpha$$

PROOF. We define a function $d\colon F^{n \times n} \to F$ by $d(\alpha) = \det \alpha^{\mathsf{T}}$ and we show that d is a determinant function. We first consider d as a function strictly of the kth row of α. But this means that we are studying $\det \alpha^{\mathsf{T}}$ as a function of its kth column. Since we know by the preceding lemma that $\det$ as a function of any particular column is a linear transformation, we conclude that d satisfies (i) of Lemma 16.1.

Suppose now that two rows of α are identical. Then two columns of α^{T} are the same and hence $0 = \det \alpha^{\mathsf{T}} = d(\alpha)$ again by the preceding lemma. Finally $d(I_n) = \det I_n^{\mathsf{T}} = \det I_n = 1$ so by the uniqueness Lemma 16.1, we have $\det \alpha^{\mathsf{T}} = d(\alpha) = \det \alpha$ and the result follows. $\square$

COROLLARY 16.1. $\det \alpha$ *can be evaluated by cofactors with respect to the rth row as follows. If $\alpha = [a_{ij}]$ then*

$$\det \alpha = \sum_{j=1}^{n} (-1)^{r+j} a_{rj} \det \alpha_{rj}$$

PROOF. Let $\alpha^{\mathsf{T}} = [a'_{ij}]$ so that $a'_{ij} = a_{ji}$. Clearly $(\alpha^{\mathsf{T}})_{ij} = (\alpha_{ji})^{\mathsf{T}}$ so using the rth column expansion of $\det \alpha^{\mathsf{T}}$ we obtain

$$\det \alpha = \det \alpha^{\mathsf{T}} = \sum_{j=1}^{n} (-1)^{j+r} a'_{jr} \det(\alpha^{\mathsf{T}})_{jr}$$

$$= \sum_{j=1}^{n} (-1)^{r+j} a_{rj} \det(\alpha_{rj})^{\mathsf{T}}$$

$$= \sum_{j=1}^{n} (-1)^{r+j} a_{rj} \det \alpha_{rj}$$

since $\det(\alpha_{rj})^{\mathsf{T}} = \det \alpha_{rj}$.   □

Now it is not true that $\det(\alpha + \beta) = \det \alpha + \det \beta$ in general. For example if

$$\alpha = \begin{bmatrix} 1 & 0 \\ 0 & 0 \end{bmatrix}, \qquad \beta = \begin{bmatrix} 0 & 0 \\ 0 & 1 \end{bmatrix}$$

then $\det \alpha = \det \beta = 0$ but $\det(\alpha + \beta) = \det I_2 = 1$. On the other hand, determinants do act well with respect to multiplication.

THEOREM 16.2. *Let $\alpha, \beta \in F^{n \times n}$. Then*

$$\det \alpha\beta = (\det \alpha)(\det \beta)$$

PROOF. Fix β throughout this proof and define $d \colon F^{n \times n} \to F$ by $d(\alpha) = \det \alpha\beta$. We observe now the expected properties of d. Note first that the kth row of $\alpha\beta$ depends on the kth row of α and all of β. Thus if α and α' agree on all rows but the kth, then so do $\alpha\beta$ and $\alpha'\beta$.

Suppose $\alpha = [a_{ij}]$, $\beta = [b_{ij}]$ and $\alpha\beta = [c_{ij}]$. Then

$$c_{kj} = \sum_{i=1}^{n} a_{ki} b_{ij}$$

so the map T from the kth row of α to the kth row of $\alpha\beta$ is given by

$$(a_{k1}, a_{k2}, \ldots, a_{kn})T = \left(\sum_{i=1}^{n} a_{ki} b_{i1}, \sum_{i=1}^{n} a_{ki} b_{i2}, \ldots, \sum_{i=1}^{n} a_{ki} b_{in} \right)$$

This is clearly a linear transformation from F^n to F^n. Thus the map $d(\alpha)$ viewed strictly as a function of the kth row of α is the composition of T followed by det viewed as a function of the kth row of $\alpha\beta$. Since the product of linear transformations is a linear transformation, we see that d satisfies (i) of Lemma 16.1.

Suppose now that two rows of α are identical. Then the corresponding two rows of $\alpha\beta$ will be identical so $d(\alpha) = \det \alpha\beta = 0$. Finally $d(I_n) = \det I_n\beta = \det \beta$ so by the uniqueness Lemma 16.1 we get

$$\det \alpha\beta = d(\alpha) = (\det \alpha){\cdot}d(I_n) = (\det \alpha)(\det \beta)$$

and the theorem is proved.   □

In the proof below, the commutativity of multiplication in F is clearly crucial.

COROLLARY 16.2. *Let* $\alpha, \beta \in F^{n\times n}$ *be similar matrices. Then* $\det \alpha = \det \beta$.

PROOF. By definition, there exists a nonsingular $n \times n$ matrix γ with $\alpha = \gamma^{-1}\beta\gamma$. Then

$$\det \alpha = \det(\gamma^{-1}\beta)\gamma = (\det \gamma^{-1}\beta)(\det \gamma)$$
$$= (\det \gamma)(\det \gamma^{-1}\beta) = \det \gamma(\gamma^{-1}\beta)$$
$$= \det \beta$$

and the result follows.   □

Finally we consider the determinant of certain block matrices. Suppose $\alpha \in F^{n\times n}$, $\beta \in F^{m\times m}$ and $\delta \in F^{m\times n}$. Then we can form the $(m + n) \times (m + n)$ block matrix

$$\gamma = \left[\begin{array}{c|c} \alpha & 0 \\ \hline \delta & \beta \end{array}\right]$$

This is of course only a symbolic expression for the matrix. It means for example that the upper left part of γ has entries identical to those of α. Observe that the upper right part of γ has all zero entries. We wish to compute $\det \gamma$ and we start with a special case.

LEMMA 16.3. *If*

$$\gamma = \left[\begin{array}{c|c} I_n & 0 \\ \hline \delta & \beta \end{array}\right]$$

then $\det \gamma = \det \beta$.

PROOF. We proceed by induction on n. If $n = 0$, then I_n and δ do not occur. Thus $\gamma = \beta$ so certainly $\det \gamma = \det \beta$. Now suppose that $n \geq 1$ and compute $\det \gamma$ by its cofactor expansion with respect to the first row. Since this row has only one nonzero entry, namely a 1 in the $1, 1$-position, we get $\det \gamma = \det \gamma_{11}$. But

$$\gamma_{11} = \left[\begin{array}{c|c} I_{n-1} & 0 \\ \hline \delta^* & \beta \end{array} \right]$$

where δ^* denotes δ with its first column deleted. By induction, $\det \gamma_{11} = \det \beta$, so the result follows. $\square$

We can now prove

THEOREM 16.3. *If*

$$\gamma = \left[\begin{array}{c|c} \alpha & 0 \\ \hline \delta & \beta \end{array} \right]$$

where $\alpha \in F^{n \times n}$, $\beta \in F^{m \times m}$ *and* $\delta \in F^{m \times n}$, *then*

$$\det \gamma = (\det \alpha)(\det \beta)$$

PROOF. Fix β and δ and define $d \colon F^{n \times n} \to F$ by $d(\alpha) = \det \gamma$ where γ is as above. Because of the presence of 0s in the upper right part of γ, we see easily that $d(\alpha)$ is a linear transformation when viewed as a function of any particular row of α. Secondly, if two rows of α are identical, then the corresponding rows of γ are the same so $d(\alpha) = \det \gamma = 0$. Thus by the uniqueness Lemma 16.1 and the preceding lemma we have

$$\det \gamma = d(\alpha) = (\det \alpha) \cdot d(I_n) = (\det \alpha)(\det \beta)$$

and the theorem is proved. $\square$

The cofactor expansion is useful theoretically, but it not a particularly efficient method computationally. Indeed it is better to use a mixture of elementary row and column operations, along with these expansions, to evaluate any fixed determinant. To start with, we have

LEMMA 16.4. *Let* $\alpha \in F^{n \times n}$ *and set* $\beta = \mathfrak{E}(\alpha)$, *where* $\mathfrak{E}$ *is an elementary row or column operation.*

    i. *If* $\mathfrak{E}$ *interchanges two distinct rows or two distinct columns, then* $\det \beta = -\det \alpha$.

ii. *If $\mathfrak{E}$ multiplies a row or a column of α by $c \in F$, then $\det \beta = c \cdot \det \alpha$.*

iii. *If $\mathfrak{E}$ adds a scalar multiple of one row to another or a scalar multiple of one column to another, then $\det \beta = \det \alpha$.*

PROOF. We first quickly consider the row operations. Note that Corollary 15.1 implies that the determinant map defines a volume function on the rows of α. Thus (i) and (iii) for row operations follow from Lemma 14.1. Furthermore, (ii) is immediate since $\det \alpha$ is a linear functional on the individual rows of α. Of course, the analogous column results follow from the fact that $\det \alpha = \det \alpha^{\mathsf{T}}$. □

It is convenient to view (ii) above as a common factor principle. Namely, if all entries in a particular row or column of α are divisible by $c \in F$, then $\det \alpha$ is equal to c times the determinant of the modified matrix with c factored out of that row or column. We now consider two examples, one rather trivial and the other rather amazing.

EXAMPLE 16.1. Let α be the 3×3 matrix

$$\alpha = \begin{bmatrix} 0 & 3 & 2 \\ 3 & 6 & 9 \\ 2 & 5 & 9 \end{bmatrix}$$

with integer entries. To evaluate $\det \alpha$, we first factor 3 out of the second row so that $\det \alpha = 3 \cdot \det \beta$ where

$$\beta = \begin{bmatrix} 0 & 3 & 2 \\ 1 & 2 & 3 \\ 2 & 5 & 9 \end{bmatrix}$$

Next, we subtract twice the second row of β from the third, so that $\det \beta = \det \gamma$, where

$$\gamma = \begin{bmatrix} 0 & 3 & 2 \\ 1 & 2 & 3 \\ 0 & 1 & 3 \end{bmatrix}$$

Finally, we evaluate $\det \gamma$ by cofactors with respect to the first column to obtain

$$\det \alpha = 3 \cdot \det \beta = 3 \cdot \det \gamma = 3 \cdot (-1)^{1+2} \cdot 1 \cdot \begin{vmatrix} 3 & 2 \\ 1 & 3 \end{vmatrix}$$

$$= 3 \cdot (-1) \cdot 7 = -21$$

Now for our more serious example.

EXAMPLE 16.2. The $n \times n$ *Hilbert Matrix* $A_n = [a_{ij}]$ with rational entries

$$a_{ij} = \frac{1}{i+j-1}$$

occurs in the study of orthogonal polynomials defined on the unit interval of the real line. One can evaluate $\det A_n$ using the theory of such polynomials or by elementary row and column operations. We of course take the latter approach and obtain the rather remarkable formula

$$\det A_n = \frac{(n-1)!!^4}{(2n-1)!!}$$

where as usual $m! = \prod_{i=1}^{m} i$, and where we set $m!! = \prod_{i=1}^{m} i!$. We proceed by induction on n and note that this formula holds for $n = 1$ since, by convention, $0! = 1$. As we see, the computations involved are simple, but a bit tedious.

Now let $n > 1$ and we apply elementary row operations to put 0 entries in the first $n - 1$ spots of the last column. Specifically, since $a_{nn} = 1/(2n-1)$, we subtract $(2n-1)/(n+i-1)$ times the nth row from the ith for $i = 1, 2, \ldots, n - 1$. This of course does not change the determinant, by Lemma 16.4(iii), but it does transform the matrix A_n to the block matrix

$$B'_n = \left[\begin{array}{c|c} B_{n-1} & 0 \\ \hline C & a_{nn} \end{array} \right]$$

where $B_{n-1} = [b_{ij}]$ and

$$b_{ij} - a_{ij} \frac{2n-1}{n+i-1} \cdot a_{nj}$$

Thus, by the cofactor expansion with respect to the last column of B'_n, we have

$$\det A_n = \det B'_n = (-1)^{n+n} a_{nn} \cdot \det B_{n-1}$$

$$= \frac{1}{2n-1} \cdot \det B_{n-1}$$

We consider the submatrix B_{n-1}. Here

$$b_{ij} = a_{ij} - \frac{2n-1}{n+i-1} \cdot a_{nj}$$

$$= \frac{1}{i+j-1} - \frac{2n-1}{n+i-1} \cdot \frac{1}{n+j-1}$$

$$= \frac{(n+i-1)(n+j-1) - (2n-1)(i+j-1)}{(n+i-1)(n+j-1)(i+j-1)}$$

$$= \frac{n^2 - n(i+j) + ij}{(n+i-1)(n+j-1)(i+j-1)}$$

$$= \frac{n-i}{n+i-1} \cdot \frac{n-j}{n+j-1} \cdot a_{ij}$$

In particular, if we factor $(n-i)/(n+i-1)$ out of the ith row of B_{n-1} for all i and $(n-j)/(n+j-1)$ out of the jth column of B_{n-1} for all j, we obtain the matrix A_{n-1}. Thus

$$\det B_{n-1} = \prod_{i=1}^{n-1} \frac{n-i}{n+i-1} \cdot \prod_{j=1}^{n-1} \frac{n-j}{n+j-1} \cdot \det A_{n-1}$$

Now

$$\prod_{i=1}^{n-1} (n-i) = (n-1)!$$

and

$$\prod_{i=1}^{n-1} (n+i-1) = \frac{(2n-2)!}{(n-1)!}$$

so

$$\det A_n = \frac{1}{2n-1} \cdot \det B_{n-1}$$

$$= \frac{1}{2n-1} \cdot \frac{(n-1)!^4}{(2n-2)!^2} \cdot \det A_{n-1}$$

$$= \frac{(n-1)!^4}{(2n-1)!(2n-2)!} \cdot \det A_{n-1}$$

Thus, by induction,

$$\det A_n = \frac{(n-1)!^4}{(2n-1)!(2n-2)!} \cdot \frac{(n-2)!!^4}{(2n-3)!!}$$

$$= \frac{(n-1)!!^4}{(2n-1)!!}$$

as required.

Problems

16.1. Show that the 2×2 determinant

$$\begin{vmatrix} a_{11} & a_{12} \\ a_{21} & a_{22} \end{vmatrix}$$

is given by the product $a_{11}a_{22}$, corresponding to the line slanting down and to the right, minus the product $a_{21}a_{12}$, corresponding to the line slanting up and to the right.

16.2. We learn this trick in calculus class for evaluating the 3×3 determinant $|a_{ij}|$. Copy the first two columns of the matrix to the right hand side, as indicated, and then draw the six slanting lines.

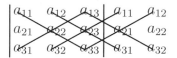

$$\left.\begin{matrix} a_{11} & a_{12} & a_{13} \\ a_{21} & a_{22} & a_{23} \\ a_{31} & a_{32} & a_{33} \end{matrix}\right| \begin{matrix} a_{11} & a_{12} \\ a_{21} & a_{22} \\ a_{31} & a_{32} \end{matrix}$$

Show that $|a_{ij}|$ is equal to the sum of the three products corresponding to the lines slanting down and to the right, minus the three products corresponding to the lines slanting up and to the right.

16.3. Can the above trick work in general for $n \times n$ determinants with $n \geq 4$? For this, consider Theorem 15.3.

16.4. If $a, b, c \in F$, show that

$$\begin{vmatrix} 1 & 1 \\ a & b \end{vmatrix} = b - a, \quad \text{and} \quad \begin{vmatrix} 1 & 1 & 1 \\ a & b & c \\ a^2 & b^2 & c^2 \end{vmatrix} = (c-a)(c-b)(b-a)$$

Why do you think these right hand factors occur?

16.5. Let $\alpha, \beta \in F^{2 \times 2}$ with

$$\alpha = \begin{bmatrix} a & b \\ c & d \end{bmatrix}, \qquad \beta = \begin{bmatrix} e & f \\ g & h \end{bmatrix}$$

Compute $\alpha\beta$ and then $\det \alpha\beta$. Prove directly that $\det \alpha\beta$ factors as the product $(\det \alpha)(\det \beta)$.

16.6. Let α and β be $n \times n$ matrices with integer entries and assume that $\alpha\beta = \text{diag}(1, 2, \ldots, n)$. Show that $\det \alpha$ is an integer dividing $n!$.

16.7. Compute the determinant

$$\begin{vmatrix} 1 & 1 & 2 & 3 \\ 0 & 2 & 4 & 5 \\ -1 & 3 & -2 & 1 \\ 2 & 4 & 1 & 9 \end{vmatrix}$$

16.8. Let $\alpha = [a_{ij}] \in F^{3 \times 3}$ and let $b_1, b_2, b_3 \in F$. Then the sum

$$(-1)^{1+2} b_1 \cdot \alpha_{12} + (-1)^{2+2} b_2 \cdot \alpha_{22} + (-1)^{3+2} b_3 \cdot \alpha_{32}$$

is the cofactor expansion with respect to the second column of some 3×3 matrix β. Describe β and determine the sum when $b_1 = a_{11}$, $b_2 = a_{21}$ and $b_3 = a_{31}$.

As in Example 16.2, let A_n be the $n \times n$ Hilbert Matrix and write $\det A_n = 1/a_n$.

16.9. Use the fact that the binomial coefficient $\binom{m}{i}$ is an integer, for all m and i, to prove that a_n is always an integer.

16.10. Verify the inequality

$$2^{2m} \geq \binom{2m}{m} \geq \frac{2^{2m}}{2m+1}$$

for all $m \geq 0$, and use this to estimate the size of the integer a_n. It clearly grows very quickly.

17. Adjoints and Inverses

Let $\alpha \in F^{n \times n}$ with $n \geq 2$. Since the subdeterminants $\det \alpha_{ij}$ are double subscripted, it is tempting to place them inside an $n \times n$ matrix. To be more precise, we form the *matrix of cofactors* $[c_{ij}]$ where $c_{ij} = (-1)^{i+j} \det \alpha_{ij}$ and then we form the *adjoint* of α given by

$$\operatorname{adj} \alpha = [a'_{ij}] = [c_{ij}]^\mathsf{T}$$

In other words, the entries of $\operatorname{adj} \alpha$ satisfy $a'_{ij} = c_{ji} = (-1)^{i+j} \det \alpha_{ji}$. Now it turns out that the row and column cofactor expansions for suitable determinants translate to yield the following remarkable result.

THEOREM 17.1. *If $\alpha \in F^{n \times n}$ with $n \geq 2$, then*

$$\alpha \cdot (\operatorname{adj} \alpha) = (\operatorname{adj} \alpha) \cdot \alpha = (\det \alpha) \cdot I_n$$

PROOF. Let $\alpha = [a_{ij}]$, $\operatorname{adj} \alpha = [a'_{ij}]$ and set $(\operatorname{adj} \alpha) \cdot \alpha = [d_{sr}]$. Then for fix subscripts r, s we have

$$d_{sr} = \sum_{i=1}^{n} a'_{si} a_{ir} = \sum_{i=1}^{n} (-1)^{i+s} a_{ir} \det \alpha_{is}$$

Now this looks like the cofactor expansion down the sth column for the determinant of some matrix β. But what is this matrix? The answer is that $\beta = [b_{ij}]$ is the matrix obtained from α by replacing the sth column of α with the rth column of α. Then β and α agree on all columns but the sth, so $\beta_{is} = \alpha_{is}$. Furthermore, $b_{is} = a_{ir}$ so Theorem 15.2 implies that

$$d_{sr} = \sum_{i=1}^{n} (-1)^{i+s} b_{is} \det \beta_{is} = \det \beta$$

Now if $r = s$, then $\beta = \alpha$ and $d_{rr} = \det \alpha$. On the other hand, if $r \neq s$, then the rth and sth columns of β are identical, so $\det \beta = 0$ and $d_{sr} = 0$. It follows that $(\operatorname{adj} \alpha) \cdot \alpha = [d_{sr}] = (\det \alpha) \cdot I_n$, as required.

The proof that $\alpha \cdot (\operatorname{adj} \alpha) = (\det \alpha) \cdot I_n$ is similar but depends upon the row cofactor expansions of certain determinants as described in Corollary 16.1. $\qquad\square$

Of course, $\operatorname{adj} \alpha$ makes no sense for $n = 1$, but we could set $\operatorname{adj} \alpha = I_n$ in this situation. This would salvage the above result in the trivial $n = 1$ case, but there is really nothing to be gained by doing so.

As a consequence of the preceding theorem, we obtain a concrete description of matrix inverses. Furthermore, we show that if a right or left inverse exists, then a two-sided inverse exists and all of these inverses are equal.

COROLLARY 17.1. *Let $\alpha \in F^{n\times n}$ with $n \geq 2$. The following are equivalent.*

> *i. There exists $\beta \in F^{n\times n}$ with $\alpha\beta = I_n$.*
> *ii. There exists $\gamma \in F^{n\times n}$ with $\gamma\alpha = I_n$.*
> *iii. $\det \alpha \neq 0$.*
> *iv. There exists $\alpha^{-1} \in F^{n\times n}$ with $\alpha\alpha^{-1} = \alpha^{-1}\alpha = I_n$.*

Furthermore, when these occur then

$$\beta = \gamma = \alpha^{-1} = \frac{1}{\det \alpha}\cdot \operatorname{adj}\alpha$$

PROOF. If β exists, then $(\det\alpha)(\det\beta) = \det I_n = 1$ and hence $\det\alpha \neq 0$. Thus (i) implies (iii), and similarly (ii) implies (iii). Next, if $\det\alpha \neq 0$, we can set

$$\alpha^{-1} = \frac{1}{\det\alpha}\cdot \operatorname{adj}\alpha$$

Then Theorem 17.1 implies that $\alpha\alpha^{-1} = \alpha^{-1}\alpha = I_n$ and (iv) holds. Of course, (iv) implies (i) and (ii).

Finally, if β is a right or two-sided inverse of α, then

$$\beta = I_n\beta = (\alpha^{-1}\alpha)\beta = \alpha^{-1}(\alpha\beta) = \alpha^{-1}I_n = \alpha^{-1}$$

Similarly, if γ exists, then $\gamma = \alpha^{-1}$, as required. $\square$

For convenience, we add a few more items to the above list of equivalent statements. Recall that the rank of a matrix is the dimension of its row space and also the dimension of its column space.

LEMMA 17.1. *Let $\alpha \in F^{n\times n}$ and let α' be the unique reduced row echelon matrix that is row equivalent to α. Then $\alpha' = I_n$ if and only if $\det\alpha \neq 0$ if and only if $\operatorname{rank}\alpha = n$.*

PROOF. It follows from Theorem 6.1 that $\operatorname{rank}\alpha = \operatorname{rank}\alpha'$ and that this rank is equal to the number of nonzero rows of α'. Furthermore, it is an easy consequence of Lemma 16.4 that $\det\alpha \neq 0$ if and only if $\det\alpha' \neq 0$. Thus we can now assume that $\alpha = \alpha'$ is in reduced row echelon form.

If $\alpha = I_n$, then certainly $\operatorname{rank}\alpha = n$ and $\det\alpha = 1 \neq 0$. Conversely, if either $\operatorname{rank}\alpha = n$ or $\det\alpha \neq 0$, then the last row of α cannot be zero. It follows that all rows of α are nonzero and therefore α has precisely n leading 1s. Since these leading 1s slope down and to the right and since α has n columns, we see that the leading 1s must fill the main diagonal and hence $\alpha = I_n$. $\square$

Since $(\alpha^{-1})^{-1} = \alpha$, one might guess that $\mathrm{adj}(\mathrm{adj}\,\alpha)$ is equal to α up to a scalar factor. However this is not quite the case since $\mathrm{adj}\,\alpha$ can be the zero matrix even when $\alpha \neq 0$. Indeed, we have

LEMMA 17.2. *Let $\alpha \in F^{n \times n}$ with $n \geq 2$. Then $\mathrm{adj}\,\alpha \neq 0$ if and only if $\mathrm{rank}\,\alpha \geq n - 1$.*

PROOF. Suppose first that $\mathrm{rank}\,\alpha \geq n - 1$. We show that some entry of $\mathrm{adj}\,\alpha$ is not 0. From the fact that the n rows of α span the row space, we see that some $n - 1$ of these rows are part of a basis for the row space. Hence at least $n - 1$ of the rows of α are linearly independent, say these are the first $n - 1$ rows.

Now let β be the $(n - 1) \times n$ matrix obtained from α by deleting the last row. Since the $n - 1$ rows of β are linearly independent, we see that β has rank $n - 1$. Hence the column space of β has dimension $n - 1$, and there are $n - 1$ columns that form a basis for this space, say these are the first $n - 1$ columns. Since α_{nn} is obtained from β by deleting the last column, it follows that the columns of α_{nn} are linearly independent. Hence $\alpha_{nn} \in F^{(n-1) \times (n-1)}$ has rank $n-1$, and we conclude from Lemma 17.1 that $c_{nn} = \det \alpha_{nn} \neq 0$. Thus $\mathrm{adj}\,\alpha \neq 0$.

Conversely, suppose $\mathrm{rank}\,\alpha \leq n - 2$ and consider the submatrix α_{rs}. First, let β be the submatrix of α obtained by deleting its rth row. Since $\mathrm{rank}\,\alpha \leq n - 2$ and β has $n - 1$ rows of α, we see that the rows of β cannot be linearly independent. Thus $\mathrm{rank}\,\beta \leq n - 2$, so the column space of β has dimension $\leq n - 2$. Since α_{rs} is obtained from β by deleting its sth column, it follows that the columns of α_{rs} cannot be linearly independent. Hence $\alpha_{rs} \in F^{(n-1) \times (n-1)}$ has rank $< n - 1$, so Lemma 17.1 implies that $\det \alpha_{rs} = 0$. Thus $\mathrm{adj}\,\alpha = 0$, as required. $\square$

As will be apparent, we can use matrix inverses to solve certain systems of linear equations. To be precise, let

$$a_{11}x_1 + a_{12}x_2 + \cdots + a_{1n}x_n = b_1$$
$$a_{21}x_1 + a_{22}x_2 + \cdots + a_{2n}x_n = b_2$$
$$\cdots\cdots\cdots$$
$$a_{n1}x_1 + a_{n2}x_2 + \cdots + a_{nn}x_n = b_n$$

be a system of n linear equations in the n unknowns $x_1, x_2, \ldots, x_n$. As usual, let $A = [a_{ij}] \in F^{n \times n}$ be the matrix of coefficients, and let

$$X = \begin{bmatrix} x_1 \\ x_2 \\ \vdots \\ x_n \end{bmatrix}, \qquad B = \begin{bmatrix} b_1 \\ b_2 \\ \vdots \\ b_n \end{bmatrix}$$

be the matrices of unknowns and constants, respectively, with these matrices both in $F^{n \times 1}$. Then the system of equations is clearly equivalent to the matrix equation

$$AX = B$$

Now if A is nonsingular, that is if the two-sided inverse A^{-1} exists, then $AX = B$ implies that

$$X = I_n X = (A^{-1}A)X = A^{-1}(AX) = A^{-1}B$$

so X can only equal $A^{-1}B$. Conversely, if $X = A^{-1}B$, then

$$AX = A(A^{-1}B) = (AA^{-1})B = I_n B = B$$

and X is indeed a solution. Notice that we use both sides of the inverse condition $A^{-1}A = AA^{-1} = I_n$ in the above. Indeed, one side yields existence and the other uniqueness.

Now we can use the formula for A^{-1} given in Corollary 17.1 along with certain cofactor expansions to obtain the following result, known as *Cramer's Rule*. However, we offer a different proof.

THEOREM 17.2. *Consider the system of n linear equations in the n unknowns $x_1, x_2, \ldots, x_n$ described by the matrix equation $AX = B$. If A is a nonsingular matrix, then*

$$x_i = \frac{\det B_i}{\det A}, \quad \text{for } i = 1, 2, \ldots, n$$

where B_i is the matrix obtained from A by replacing its ith column by the column of B.

PROOF. Since A is nonsingular, we know that $X \in F^{n \times 1}$ is uniquely determined by this system of equations. Furthermore, $\det A \neq 0$. Let $x_1, x_2, \ldots, x_n \in F$ be the solution set and consider x_i. Then $x_i \cdot \det A = \det A_i$, where A_i is obtained from A by multiplying its ith column by x_i. Next, for all $k \neq i$, we add x_k times the kth column of A_i to the ith column. This does not change the determinant, but it does transform the matrix A_i into the matrix B_i. Indeed, the jth entry of the ith column of this new matrix is

$$a_{ji}x_i + \sum_{k \neq i} a_{jk}x_k = b_j$$

Thus $x_i \cdot \det A = \det A_i = \det B_i$ and the lemma is proved since we know that $\det A \neq 0$. $\qquad \square$

This result is of interest for two reasons. First, if A and B are integer matrices, then Cramer's Rule gives us an idea of the possible denominators that can show up in solutions. Second, unlike the usual approach to solving linear equations, we can find one unknown here, without having to find them all. A simple example is as follows.

EXAMPLE 17.1. Consider the system of 3 linear equations in the 3 unknowns x_1, x_2, x_3 with corresponding matrices

$$A = \begin{bmatrix} 1 & 1 & 0 \\ 2 & 3 & 1 \\ 1 & 2 & 3 \end{bmatrix}, \qquad B = \begin{bmatrix} 2 \\ 8 \\ 9 \end{bmatrix}$$

Then $\det A = 2$ and

$$B_1 = \begin{bmatrix} 2 & 1 & 0 \\ 8 & 3 & 1 \\ 9 & 2 & 3 \end{bmatrix}, \qquad B_2 = \begin{bmatrix} 1 & 2 & 0 \\ 2 & 8 & 1 \\ 1 & 9 & 3 \end{bmatrix}, \qquad B_3 = \begin{bmatrix} 1 & 1 & 2 \\ 2 & 3 & 8 \\ 1 & 2 & 9 \end{bmatrix}$$

Thus we have $x_1 = \det B_1 / \det A = -1/2$, $x_2 = \det B_2 / \det A = 5/2$ and $x_3 = \det B_3 / \det A = 3/2$.

Now it is tedious enough finding one $n \times n$ determinant, so evaluating n^2 determinants is just too much to ask for. Thus we really cannot use the adjoint formula to compute the inverse of a large square matrix. Fortunately, there is a simple computational technique that does the job rather efficiently.

Let $A \in F^{n \times n}$ and form the $n \times (2n)$ augmented matrix $A|I_n$. If A is nonsingular, then A is row equivalent to the identity matrix I_n, and using the same operations we see that $A|I_n$ is row equivalent to $I_n|B$ for some matrix $B \in F^{n \times n}$. Furthermore, $I_n|B$ is in reduced row echelon form, so $I_n|B$ and hence B are uniquely determined by this process. Indeed, it turns out that $B = A^{-1}$. We consider some examples.

EXAMPLE 17.2. Let

$$A = \begin{bmatrix} 1 & 0 & 1 \\ 1 & 1 & 2 \\ 2 & 1 & 3 \end{bmatrix}$$

Then the augmented matrix is given by

$$A|I_3 = \left[\begin{array}{ccc|ccc} 1 & 0 & 1 & 1 & 0 & 0 \\ 1 & 1 & 2 & 0 & 1 & 0 \\ 2 & 1 & 3 & 0 & 0 & 1 \end{array} \right]$$

and we now apply elementary row operations. To start with, subtract the first row from the second and subtract twice the first row from the third. This yields

$$\left[\begin{array}{ccc|ccc} 1 & 0 & 1 & 1 & 0 & 0 \\ 0 & 1 & 1 & -1 & 1 & 0 \\ 0 & 1 & 1 & -2 & 0 & 1 \end{array}\right]$$

Next, subtracting row 2 from row 3, we get

$$\left[\begin{array}{ccc|ccc} 1 & 0 & 1 & 1 & 0 & 0 \\ 0 & 1 & 1 & -1 & 1 & 0 \\ 0 & 0 & 0 & -1 & -1 & 1 \end{array}\right]$$

and this makes no sense. The three zeroes in the last row tell us that A must have been a singular matrix, so A^{-1} does not exist.

EXAMPLE 17.3. Let us start again, this time with

$$A = \left[\begin{array}{ccc} 1 & 0 & 1 \\ 1 & 1 & 2 \\ 2 & 1 & 4 \end{array}\right]$$

so that

$$A|I_3 = \left[\begin{array}{ccc|ccc} 1 & 0 & 1 & 1 & 0 & 0 \\ 1 & 1 & 2 & 0 & 1 & 0 \\ 2 & 1 & 4 & 0 & 0 & 1 \end{array}\right]$$

Again, we subtract the first row from the second and subtract twice the first row from the third. This yields

$$\left[\begin{array}{ccc|ccc} 1 & 0 & 1 & 1 & 0 & 0 \\ 0 & 1 & 1 & -1 & 1 & 0 \\ 0 & 1 & 2 & -2 & 0 & 1 \end{array}\right]$$

Next, subtracting row 2 from row 3, gives us

$$\left[\begin{array}{ccc|ccc} 1 & 0 & 1 & 1 & 0 & 0 \\ 0 & 1 & 1 & -1 & 1 & 0 \\ 0 & 0 & 1 & -1 & -1 & 1 \end{array}\right]$$

and finally subtracting row 3 from rows 1 and 2 yields

$$\left[\begin{array}{ccc|ccc} 1 & 0 & 0 & 2 & 1 & -1 \\ 0 & 1 & 0 & 0 & 2 & -1 \\ 0 & 0 & 1 & -1 & -1 & 1 \end{array}\right] = I_3|B$$

Thus

$$B = \begin{bmatrix} 2 & 1 & -1 \\ 0 & 2 & -1 \\ -1 & -1 & 1 \end{bmatrix}$$

must equal A^{-1} and we can check this by verifying that $AB = BA = I_3$.

We state this procedure as a formal lemma and offer a proof.

LEMMA 17.3. *If $A \in F^{n \times n}$ is nonsingular, then the augmented matrix $A|I_n$ is row equivalent to the reduced row echelon matrix $I_n|A^{-1}$.*

FIRST PROOF. To avoid the subscript on the identity matrix, let us write $I_n = J$. We consider the matrix equation $AX = J$, where X has size $n \times n$. A solution here is a right inverse for A and hence the unique two-sided inverse A^{-1} by Corollary 17.1. Notice that the jth column of X, namely the column matrix X_j, determines the jth column of the product $AX = J$, namely the column matrix J_j. Thus $AX = J$ is equivalent to the n matrix equations $AX_j = J_j$ for $j = 1, 2, \ldots, n$.

Now, for each j, we solve $AX_j = J_j$ by using the techniques of Section 7. Namely, we first form the augmented matrix $A|J_j$ and then put this into reduced row echelon form. Since A is row equivalent to $I_n = J$, the same elementary row operations will transform $A|J_j$ to $J|B_j$ for some column matrix B_j and this system of equations has the same solution set as the original by Lemma 7.1. Writing this new system as a matrix product yields $B_j = JX_j = X_j$.

Of course, we can solve all n systems simultaneously by forming the extended augmentation matrix

$$A|J_1|J_2| \cdots |J_n = A|J$$

and applying the same row operations that transform A into J. We see that $A|J$ is transformed to

$$J|B_1|B_2| \cdots |B_n = J|B$$

where B_j is the jth column of $B \in F^{n \times n}$. But $B_j = X_j$ for all j, so $B = X = A^{-1}$, as required.

Finally, note that $J|B$ is in reduced row echelon form, so it is the unique reduced row echelon form matrix obtainable from $A|J$. Hence it is independent of the particular sequence of elementary row operations that are used. $\qquad\square$

Notice that the proof above uses a right inverse for A. There is another proof, interesting in its own right, that uses a left inverse for A. To start with, we have a result we have seen before.

LEMMA 17.4. *Let $A \in F^{m \times n}$ and $B \in F^{n \times r}$, so that $AB \in F^{m \times r}$. If $\mathfrak{E}$ is an elementary row operation, then $\mathfrak{E}(AB) = \mathfrak{E}(A) \cdot B$. In particular, $\mathfrak{E}(B) = \mathfrak{E}(I_n) \cdot B$.*

PROOF. As we know, the rth row of AB is determined by the rth row of A and all entries of B. In particular, if we interchange two rows of A, then the two corresponding rows of AB are interchanged. Furthermore, if we multiply the rth row of A by $c \in F$, then the rth row of AB is also multiplied by c. The third elementary row operation can also be visualized, but we offer a more concrete argument. Let $A = [a_{ij}]$, $B = [b_{ij}]$ and $AB = [c_{ij}]$. If we add x times the rth row of A to the sth row, then a_{sj} becomes $a'_{sj} = a_{sj} + x \cdot a_{rj}$. Thus c_{sj} becomes

$$c'_{sj} = \sum_k a'_{sk} b_{kj} = \sum_k (a_{sk} + x \cdot a_{rk}) b_{kj}$$

$$= \sum_k a_{sk} b_{kj} + x \cdot \sum_k a_{rk} b_{kj} = c_{sj} + x \cdot c_{rj}$$

and this says that the new sth row of AB is the old sth row plus x times the rth row.

Finally, since $B = I_n B$, we see that $\mathfrak{E}(B) = \mathfrak{E}(I_n B) = \mathfrak{E}(I_n) B$ and the lemma is proved. $\qquad\square$

The various $\mathfrak{E}(I_n)$ above are called *elementary matrices*. As a consequence, we have the following second proof of Lemma 17.3.

SECOND PROOF. Since $A \in F^{n \times n}$ is nonsingular, it is row equivalent to I_n. Thus there exists a sequence $\mathfrak{E}_1, \mathfrak{E}_2, \ldots, \mathfrak{E}_k$ of elementary row operations that transform A to I_n. In particular, if $E_i = \mathfrak{E}_i(I_n)$, then the preceding lemma implies that

$$E_k \cdots E_2 E_1 \cdot A = \mathfrak{E}_k \cdots \mathfrak{E}_2 \mathfrak{E}_1(A) = I_n$$

and thus $E_k \cdots E_2 E_1$ is a left inverse for A. By Corollary 17.1, it is the unique two-sided inverse, namely A^{-1}. But again, by Lemma 17.4,

$$E_k \cdots E_2 E_1 = E_k \cdots E_2 E_1 \cdot I_n = \mathfrak{E}_k \cdots \mathfrak{E}_2 \mathfrak{E}_1(I_n)$$

In other words, the elementary row operations that transform A into I_n, also transform I_n into A^{-1} and this is precisely what the process of Lemma 17.3 asserts. Indeed by considering the augmented matrix $A|I_n$, we apply the same row operations to A and to I_n, in the same order, so when $A|I_n$ becomes $I_n|B$, we must have $B = A^{-1}$. $\qquad\square$

Problems

17.1. Let $\alpha \in F^{n \times n}$ with $n \geq 2$. If $\det \alpha = 0$, show that $\det(\operatorname{adj} \alpha) = 0$. For this, note that if $\det(\operatorname{adj} \alpha) \neq 0$, then $\operatorname{adj} \alpha$ is invertible, and then $(\operatorname{adj} \alpha)\alpha = 0$ implies that $\alpha = 0$. As a consequence, conclude that for all α, we have $\det(\operatorname{adj} \alpha) = (\det \alpha)^{n-1}$.

17.2. If α, β are nonsingular matrices in $F^{n \times n}$, show that $(\alpha\beta)^{-1} = \beta^{-1}\alpha^{-1}$ and then that $\operatorname{adj}(\alpha\beta) = (\operatorname{adj} \beta)(\operatorname{adj} \alpha)$.

17.3. For
$$\alpha = \begin{bmatrix} a & b \\ c & d \end{bmatrix}$$
find $\operatorname{adj} \alpha$ and $\operatorname{adj}(\operatorname{adj} \alpha)$. If α is nonsingular, find α^{-1}.

17.4. If $\alpha \in F^{n \times n}$ has rank $n - 1$, use Problem 12.3 to prove that $\operatorname{adj} \alpha$ has rank 1. In particular, if $n \geq 3$, conclude that $\operatorname{adj}(\operatorname{adj} \alpha) = 0$.

17.5. Compute the adjoint of the 3×3 real matrix
$$\begin{bmatrix} 1 & 1 & 0 \\ 2 & 3 & -1 \\ 1 & 4 & -1 \end{bmatrix}$$
Find the inverse of this matrix using Corollary 17.1 and also by using the technique of Lemma 17.3.

17.6. Solve the system of real linear equations
$$x_1 + x_2 + 2x_3 = 4$$
$$x_1 + 2x_2 + 2x_3 = 9$$
$$2x_1 + 3x_2 + 7x_3 = 7$$
using Cramer's Rule.

17.7. Prove Cramer's Rule using the inverse of the matrix A given by Corollary 17.1 and suitable column cofactor expansions.

17.8. State and prove the column analog of Lemma 17.4. Show that the associative law of matrix multiplication implies that any elementary row operation commutes with any elementary column operation. For this, see Problem 6.3.

17.9. Show that any elementary matrix obtained from an elementary column operation applied to I_n is equal to an elementary matrix obtained from an elementary row operation applied to I_n.

17.10. Prove that elementary matrices are all nonsingular with elementary inverses, and that any nonsingular square matrix is a product of elementary matrices.

18. The Characteristic Polynomial

Now that we know determinants exist, it is a simple matter to solve the eigenvalue problem. Let $T\colon V \to V$ be a linear transformation. If $a \in F$ is an eigenvalue for T, then the linear transformation $S = aI - T$ is singular and thus for any basis $\mathcal{B}$ of V we have

$$\det {}_{\mathcal{B}}[aI - T]_{\mathcal{B}} = \det {}_{\mathcal{B}}[S]_{\mathcal{B}} = 0$$

Therefore what we want to study is $\det {}_{\mathcal{B}}[aI - T]_{\mathcal{B}}$. Now this is clearly a function of $a \in F$ and the easiest way to indicate this fact is to replace a by some variable like x. But then, what does xI mean? Unfortunately it doesn't mean anything. We avoid this difficulty by first considering matrices.

Let F be a field and, as in Example 2.5, let $F[x]$ be the ring of polynomials over F in the variable x. Then $F[x]$ embeds in a certain field extension $F(x)$ of F, the so-called field of *rational functions*. Specifically, this is the set of all quotients of polynomials with coefficients in F. More precisely, $F(x)$ is the set of equivalence classes of such fractions. It bears the same relationship to $F[x]$ as the rational field $\mathbb{Q}$ does to the ring of integers $\mathbb{Z}$. Note that $F \subseteq F[x] \subseteq F(x)$ with F being the set of constant polynomials. With this, we have $F^{n\times n} \subseteq F(x)^{n\times n}$. In other words, $F^{n\times n}$ is the set of all $n \times n$ matrices over $F(x)$ all of whose entries lie in F.

Now F is a subfield of $F(x)$. This means that when we add or multiply two elements of F, it does not matter whether we view these elements as belonging to F or to $F(x)$. Certainly this property carries over to say that if we add or multiply two matrices in $F^{n\times n}$, it does not matter whether we view these as belonging to $F^{n\times n}$ or $F(x)^{n\times n}$. Moreover all our previous results on matrices and determinants hold for $F(x)^{n\times n}$, since after all they were proved for an arbitrary field.

Let $\alpha \in F^{n\times n}$. We define its *characteristic polynomial* $\varphi_\alpha(x)$ to be

$$\varphi_\alpha(x) = \det(xI_n - \alpha)$$

where $xI_n - \alpha$ is viewed as a matrix in $F(x)^{n\times n}$. The following lemma justifies the name. Note that a polynomial is said to be *monic* if its leading coefficient is equal to 1.

LEMMA 18.1. *If $\alpha \in F^{n\times n}$, then $\varphi_\alpha(x)$ is a monic polynomial of degree n.*

PROOF. If $\alpha = [a_{ij}]$, then $xI_n - \alpha = [b_{ij}]$ where $b_{ij} = -a_{ij}$ for $i \neq j$ and $b_{ii} = x - a_{ii}$. By Theorem 15.3, we have

$$\varphi_\alpha(x) = |b_{ij}| = \sum \pm b_{1j_1} b_{2j_2} \cdots b_{nj_n}$$

where the sum is over all n-tuples $(j_1, j_2, \ldots, j_n)$ of distinct column subscripts. Since each b_{ij} is a polynomial in x of degree ≤ 1, we see immediately that $\varphi_\alpha(x) \in F[x]$ and has degree $\leq n$. Now the only way we can get a term of degree n in this sum is for each b_{ij} to have degree 1 and this occurs only for the term

$$+b_{11}b_{22}\cdots b_{nn} = (x - a_{11})(x - a_{22})\cdots(x - a_{nn})$$
$$= x^n + \text{smaller degree terms}$$

Thus $\varphi_\alpha(x)$ has degree n with leading coefficient equal to 1. □

In order to define the characteristic polynomial of a linear transformation, we need the following.

LEMMA 18.2. *Let* $\alpha, \beta \in F^{n \times n}$ *be similar matrices. Then* $\varphi_\alpha(x) = \varphi_\beta(x)$.

PROOF. By definition, there exists a nonsingular matrix $\gamma \in F^{n \times n}$ with $\gamma^{-1}\beta\gamma = \alpha$. Let us work in $F(x)^{n \times n}$. Since it is easy to see that $(xI_n)\gamma = \gamma(xI_n) = x\gamma$, we have

$$\gamma^{-1}(xI_n - \beta)\gamma = \gamma^{-1}(xI_n)\gamma - \gamma^{-1}\beta\gamma$$
$$= \gamma^{-1}\gamma(xI_n) - \gamma^{-1}\beta\gamma = xI_n - \alpha$$

and thus $xI_n - \beta$ and $xI_n - \alpha$ are similar matrices in $F(x)^{n \times n}$. By Corollary 16.2, we conclude that

$$\varphi_\alpha(x) = \det(xI_n - \alpha) = \det(xI_n - \beta) = \varphi_\beta(x)$$

and the lemma is proved. □

Let V be an n-dimensional vector space over F and let $T\colon V \to V$ be a linear transformation. We define the *characteristic polynomial* $\varphi_T(x)$ of T to be

$$\varphi_T(x) = \varphi_\alpha(x)$$

where $\alpha = {}_\mathcal{A}[T]_\mathcal{A}$ for some basis $\mathcal{A}$ of V. Observe that if $\mathcal{B}$ is a second basis and if $\beta = {}_\mathcal{B}[T]_\mathcal{B}$, then α and β are similar matrices by Lemma 13.1(i). Now by the above, such matrices have the same characteristic polynomial, and thus we see that $\varphi_T(x)$ is well defined. In particular, we have proved

LEMMA 18.3. *Let* $T\colon V \to V$ *be a linear transformation. Then* $\varphi_T(x)$ *is a monic polynomial of degree* $n = \dim_F V$. *Moreover, this polynomial can be found by choosing any basis* $\mathcal{A}$ *for* V *and then setting* $\varphi_T(x) = \varphi_\alpha(x)$ *where* $\alpha = {}_\mathcal{A}[T]_\mathcal{A}$.

Recall that if $f(x) \in F[x]$, then we can consider f to be a function $f \colon F \to F$ by evaluation. Namely if $a \in F$, then we obtain $f(a)$ by replacing x by a in the polynomial expression for f. We have easily for $f(x), g(x) \in F[x]$ and $a \in F$

$$(f + g)(a) = f(a) + g(a)$$
$$(fg)(a) = f(a)g(a)$$

where the latter is multiplication of polynomials and not composition. If $f(a) = 0$, we say that a is a *root* of $f(x)$.

The main result connecting eigenvalues of T and the characteristic polynomial $\varphi_T(x)$ is

THEOREM 18.1. *Let V be a finite dimensional vector space over F and let $T \colon V \to V$ be a linear transformation. Then $a \in F$ is an eigenvalue for T if and only if a is a root of $\varphi_T(x)$.*

PROOF. Fix a basis $\mathcal{B}$ for V and let $a \in F$. Then we know by Theorem 13.2 that a is an eigenvalue for T if and only if the linear transformation $S = aI - T$ is singular and hence by Corollary 15.2 if and only if $\det {}_\mathcal{B}[S]_\mathcal{B} = 0$. Now

$$_\mathcal{B}[S]_\mathcal{B} = {}_\mathcal{B}[aI - T]_\mathcal{B} = {}_\mathcal{B}[aI]_\mathcal{B} - {}_\mathcal{B}[T]_\mathcal{B} = aI_n - \alpha$$

where $\alpha = {}_\mathcal{B}[T]_\mathcal{B}$. Thus a is an eigenvalue for T if and only if

$$0 = \det {}_\mathcal{B}[S]_\mathcal{B} = \det(aI_n - \alpha) = \varphi_T(a)$$

and therefore if and only if a is a root of $\varphi_T(x)$. $\square$

In order to better use the above result, we briefly consider roots of polynomials.

LEMMA 18.4. *Let $f(x) \in F[x]$ be a nonzero polynomial.*

  i. *If $a \in F$, then a is a root of $f(x)$ if and only if we have $f(x) = (x - a)g(x)$ for some polynomial $g(x) \in F[x]$.*
  ii. *$f(x)$ can be written uniquely up to the order of the factors as the product*

$$f(x) = (x - a_1)(x - a_2) \cdots (x - a_r)g(x)$$

  where $g(x) \in F[x]$ has no roots in F. Thus $\{a_1, a_2, \ldots, a_r\}$ is the set of roots of $f(x)$ in F.

PROOF. (i) If $f(x) = (x - a)g(x)$, then clearly $f(a) = (a - a)g(a) = 0$ and a is a root of $f(x)$. Conversely, suppose a is a root. By formal long division, we divide $f(x)$ by $(x - a)$ and obtain

$$f(x) = (x - a)g(x) + b$$

where b is some element of F. Setting $x = a$ in this expression yields $b = 0$ and this fact follows.

(ii) We show the existence of the required representation for $f(x)$ by induction on the degree of $f(x)$. If $f(x)$ has no roots in F, then $f(x) = f(x)$ is such a product and we are done. Now suppose $f(x)$ has a root $a_1 \in F$. By (i), $f(x) = (x - a_1)h(x)$ for some $h(x) \in F[x]$. Since clearly $\deg h(x) = \deg f(x) - 1$, we see by induction that $h(x) = (x - a_2)(x - a_3)\cdots(x - a_r)g(x)$ where $g(x)$ has no roots in F. Thus

$$f(x) = (x - a_1)h(x) = (x - a_1)(x - a_2)\cdots(x - a_r)g(x)$$

and existence is proved.

Now suppose $f(x)$ is written as above. Then clearly the elements $a_1, a_2, \ldots, a_r$ are roots of $f(x)$. Conversely suppose $a \in F$ is a root of $f(x)$. Then

$$0 = f(a) = (a - a_1)(a - a_2)\cdots(a - a_r)g(a)$$

Since $g(x)$ has no roots in F, we see that $g(a) \neq 0$ and thus we must have $a - a_i = 0$ for some i. Therefore $a = a_i$ and $\{a_1, a_2, \ldots, a_r\}$ is precisely the set of roots of $f(x)$ in F.

Finally, we show that the above expression for $f(x)$ is unique by induction on $\deg f(x)$. Suppose

$$f(x) = (x - a_1)(x - a_2)\cdots(x - a_r)g(x)$$
$$= (x - b_1)(x - b_2)\cdots(x - b_s)h(x)$$

where $g(x)$ and $h(x)$ have no roots in F. If $s = 0$, then $f(x) = h(x)$ has no roots in F. This implies that no a_is can occur, so $r = 0$ and $g(x) = f(x) = h(x)$. On the other hand, if $s > 0$, then b_1 is a root of $f(x)$, so $b_1 \in \{a_1, a_2, \ldots, a_r\}$. Say $b_1 = a_1$. Since $f(x) \in F[x] \subseteq F(x)$ and $F(x)$ is a field, we can cancel the common factor $x - a_1 = x - b_1$ and conclude that

$$(x - a_2)\cdots(x - a_r)g(x) = (x - b_2)\cdots(x - b_s)h(x)$$

By induction, uniqueness follows.   □

Suppose now that $f(x) = (x - a_1)(x - a_2)\cdots(x - a_r)g(x)$ where $g(x)$ has no roots in F. Then $f(x)$ has r roots, but they may not all be distinct. Thus, to be precise, we say that $f(x)$ has r roots *counting multiplicities*, that is counting a root b the number of times $(x - b)$ occurs as a factor of $f(x)$.

COROLLARY 18.1. *If $T: V \to V$ is a linear transformation with $\dim_F V = n$, then $\varphi_T(x)$ has at most n roots in F counting multiplicities. Thus T has at most n eigenvalues.*

If $F = \mathbb{R}$, then we know that the polynomial $f(x) = x^2 + 1$ has no roots in $\mathbb{R}$. However it does have roots in $\mathbb{C}$. In fact, the field of complex numbers has a very nice property, namely it is *algebraically closed*. Formally a field F is said to be algebraically closed if every nonconstant polynomial $f(x) \in F[x]$ has at least one root in F. There are many examples of such fields but they are difficult to construct. So we will just be content with having the one example $F = \mathbb{C}$.

LEMMA 18.5. *Let F be an algebraically closed field. If $f(x) \in F[x]$ has degree $n > 0$, then $f(x)$ has precisely n roots in F counting multiplicities.*

PROOF. Write $f(x)$ as the product

$$f(x) = (x - a_1)(x - a_2) \cdots (x - a_r)g(x)$$

where $g(x)$ has no roots in F. Since F is algebraically closed, it follows that $g(x)$ is a constant polynomial. Thus $r = n$ and $f(x)$ has n roots counting multiplicities. $\square$

COROLLARY 18.2. *Assume that F is an algebraically closed field. If $T: V \to V$ is a linear transformation and $\dim_F V = n$, then $\varphi_T(x)$ has exactly n roots in F counting multiplicities. Thus T has at least one eigenvalue and eigenvector.*

Unfortunately, the existence of these n eigenvalues, counting multiplicities, does not guarantee that the transformation T can be diagonalized. For example, if

$$_{\mathcal{A}}[T]_{\mathcal{A}} = \begin{bmatrix} 0 & 0 \\ 1 & 0 \end{bmatrix}$$

then $\varphi_T(x) = x^2$ so T has the eigenvalue 0 with multiplicity 2. If T could be diagonalized, then for some basis $\mathcal{B}$ we would have $_{\mathcal{B}}[T]_{\mathcal{B}} = \operatorname{diag}(0,0)$ and $T = 0$, a contradiction. We do however have

LEMMA 18.6. *Let $a_1, a_2, \ldots, a_k$ be distinct eigenvalues for the linear transformation T with corresponding eigenvectors $\alpha_1, \alpha_2, \ldots, \alpha_k$. Then $\{\alpha_1, \alpha_2, \ldots, \alpha_k\}$ is linearly independent.*

PROOF. We proceed by induction on k. If $k = 1$, then $\{\alpha_1\}$ is linearly independent since $\alpha_1 \neq 0$ by assumption. Suppose the result is true for $k - 1$ and let

$$0 = b_1 \alpha_1 + b_2 \alpha_2 + \cdots + b_k \alpha_k$$

Since $\alpha_i T = a_i \alpha_i$, we obtain by applying T,

$$0 = 0T = (b_1 \alpha_1 + b_2 \alpha_2 + \cdots + b_k \alpha_k)T$$
$$= b_1 a_1 \alpha_1 + b_2 a_2 \alpha_2 + \cdots + b_k a_k \alpha_k$$

Now by subtracting the second displayed equation from a_k times the first, we have

$$0 = b_1(a_k - a_1)\alpha_1 + b_2(a_k - a_2)\alpha_2 + \cdots + b_{k-1}(a_k - a_{k-1})\alpha_{k-1}$$

and thus by induction $b_i(a_k - a_i) = 0$ for $i = 1, 2, \ldots, k - 1$. Since the a_is are distinct, $a_k - a_i \neq 0$, so $b_i = 0$ for $i = 1, 2, \ldots, k - 1$. The first equation then reduces to $b_k \alpha_k = 0$ and since $\alpha_k \neq 0$ we also have $b_k = 0$. Thus $\{\alpha_1, \alpha_2, \ldots, \alpha_k\}$ is linearly independent. $\square$

As a consequence, we conclude that

THEOREM 18.2. *Let $T\colon V \to V$ be a linear transformation with V an n-dimensional vector space over F. If $\varphi_T(x)$ has n distinct roots in F, then T can be diagonalized.*

PROOF. Let the roots of $\varphi_T(x)$ be $a_1, a_2, \ldots, a_n \in F$. Then each a_i is an eigenvalue of T and let α_i be some corresponding eigenvector. Since the a_is are distinct, the preceding lemma implies that $\mathcal{A} = \{\alpha_1, \alpha_2, \ldots, \alpha_n\}$ is a linearly independent subset of V of size $n = \dim_F V$. Thus $\mathcal{A}$ is a basis for V and V has a basis consisting entirely of eigenvectors of T. This implies that T can be diagonalized and in fact $_{\mathcal{A}}[T]_{\mathcal{A}} = \mathrm{diag}(a_1, a_2, \ldots, a_n)$. $\square$

We close this section with another application of the embedding of $F^{n \times n}$ into $F(x)^{n \times n}$.

EXAMPLE 18.1. Let $a_1, a_2, \ldots, a_n$ be n elements of F and form the $n \times n$ *Vandermonde matrix*

$$\nu = \begin{bmatrix} 1 & 1 & \cdots & 1 \\ a_1 & a_2 & \cdots & a_n \\ a_1^2 & a_2^2 & \cdots & a_n^2 \\ & \cdots\cdots\cdots & \\ a_1^{n-1} & a_2^{n-1} & \cdots & a_n^{n-1} \end{bmatrix}$$

Notice that if $a_i = a_j$ for some $i \neq j$, then two columns of this matrix are equal and hence $\det \nu = 0$. We claim that in general

$$\det \nu = \prod_{i>j}(a_i - a_j)$$

Since this holds in the trivial situation when two of the parameters are equal, we can assume in the proof by induction that all a_i are distinct. Obviously this result holds for $n = 1$ and 2, so let $n > 2$.

Now work in $F(x)^{n \times n}$ and form the $n \times n$ matrix

$$\alpha = \begin{bmatrix} 1 & 1 & \cdots & 1 \\ a_1 & a_2 & \cdots & x \\ a_1^2 & a_2^2 & \cdots & x^2 \\ & & \cdots\cdots\cdots & \\ a_1^{n-1} & a_2^{n-1} & \cdots & x^{n-1} \end{bmatrix}$$

where a_n is replaced by the variable x. By considering the column cofactor expansion with respect to the nth column, we see that $\det \alpha$ is a polynomial in x of degree $\leq n - 1$. Indeed, it has degree precisely $n - 1$ since the coefficient of x^{n-1} is $c_{n-1} = \det \alpha_{nn}$, the Vandermonde determinant determined by $a_1, a_2, \ldots, a_{n-1}$. Furthermore, $c_{n-1} \neq 0$ since the parameters are distinct.

Finally note that $\det \alpha$ is 0 when evaluated at $x = a_1, a_2, \ldots, a_{n-1}$ since in each of these cases, two columns of the matrix are equal. Since these $n - 1$ roots are all distinct, we must have

$$\det \alpha = c_{n-1}(x - a_1)(x - a_2) \cdots (x - a_{n-1})$$

and evaluating at $x = a_n$ yields

$$\det \nu = c_{n-1} \prod_{n > j}(a_n - a_j) = \prod_{n \neq i > j}(a_i - a_j) \cdot \prod_{n > j}(a_n - a_j)$$

$$= \prod_{i > j}(a_i - a_j)$$

as required.

Problems

18.1. If $\alpha \in F^{n \times n}$ and $a \in F$, show that $\alpha(aI_n) = (aI_n)\alpha$.

Compute the characteristic polynomials of the following two matrices with entries in the field F.

18.2.

$$\alpha = \begin{bmatrix} a_{11} & a_{12} & a_{13} & a_{14} & a_{15} \\ 0 & a_{22} & a_{23} & a_{24} & a_{25} \\ 0 & 0 & a_{33} & a_{34} & a_{35} \\ 0 & 0 & 0 & a_{44} & a_{45} \\ 0 & 0 & 0 & 0 & a_{55} \end{bmatrix}$$

18.3.
$$\alpha = \begin{bmatrix} 0 & 1 & 0 & 0 \\ 0 & 0 & 1 & 0 \\ 0 & 0 & 0 & 1 \\ a & b & c & d \end{bmatrix}$$

If $\alpha = [a_{ij}] \in F^{n \times n}$, then the *trace* of α is defined to be
$$\operatorname{tr} \alpha = a_{11} + a_{22} + \cdots + a_{nn}$$
the sum of the entries of α on the main diagonal.

18.4. Prove that $\operatorname{tr}\colon F^{n \times n} \to F$ is a linear functional and that $\operatorname{tr}(\alpha\beta) = \operatorname{tr}(\beta\alpha)$ for all $\alpha, \beta \in F^{n \times n}$.

18.5. If $\alpha \in F^{n \times n}$, show that
$$\varphi_\alpha(x) = x^n - (\operatorname{tr} \alpha)x^{n-1} + \text{smaller degree terms}$$
Furthermore, show that the constant term of $\varphi_\alpha(x)$ is $(-1)^n \det \alpha$.

Let $\dim_F V = n$ and let $T\colon V \to V$ be a linear transformation. We define $\operatorname{tr} T$ and $\det T$ by
$$\operatorname{tr} T = \operatorname{tr} \alpha, \qquad \det T = \det \alpha$$
where $\alpha = {}_\mathcal{A}[T]_\mathcal{A}$ for some basis $\mathcal{A}$ of V.

18.6. Prove that $\operatorname{tr} T$ and $\det T$ are well defined, that is they are independent of the choice of basis $\mathcal{A}$.

18.7. Let F be algebraically closed and let $a_1, a_2, \ldots, a_n$ be the eigenvalues of T counting multiplicities. Show that
$$\operatorname{tr} T = a_1 + a_2 + \cdots + a_n, \qquad \det T = a_1 a_2 \cdots a_n$$

18.8. Let V be a 2-dimensional vector space over the real field and let $T\colon V \to V$ satisfy
$${}_\mathcal{A}[T]_\mathcal{A} = \begin{bmatrix} 1 & -1 \\ 3 & 2 \end{bmatrix}$$
Find $\varphi_T(x)$ and then using $T^0 = I$, find the matrix of the linear transformation $\psi_T(T)$.

If $\alpha \in F^{n \times n}$, define $\alpha(x) \in F(x)^{n \times n}$ by $\alpha(x) = xI_n + \alpha$.

18.9. Show that $\alpha(x)$ is a nonsingular matrix with entries in $F[x]$ and satisfying $\alpha(0) = \alpha$.

18.10. Let $\alpha, \beta \in F^{n \times n}$. Use Problem 17.2 and the above to show that $\operatorname{adj}(\alpha(x)\beta(x)) = (\operatorname{adj}\beta(x)) \cdot (\operatorname{adj}\alpha(x))$. Now evaluate at $x = 0$ to conclude that $\operatorname{adj}(\alpha\beta) = (\operatorname{adj}\beta) \cdot (\operatorname{adj}\alpha)$.

19. The Cayley-Hamilton Theorem

There are other properties of characteristic polynomials that are of interest. We start with a simple question. Is there anything really special about a characteristic polynomial? First, we know it must have degree n where $n = \dim_F V$ and second it must be monic. Are there other requirements on the polynomial? The answer is "no".

Let $f(x) \in F[x]$ be a monic polynomial of degree n so that

$$f(x) = a_0 + a_1 x + \cdots + a_{n-1}x^{n-1} + x^n$$

We define the *companion matrix* of f to be the $n \times n$ F-matrix

$$\alpha = \begin{bmatrix} 0 & 1 & 0 & 0 & \cdots & 0 & 0 \\ 0 & 0 & 1 & 0 & \cdots & 0 & 0 \\ 0 & 0 & 0 & 1 & \cdots & 0 & 0 \\ & & & \cdots\cdots\cdots\cdots\cdots & & \\ 0 & 0 & 0 & 0 & \cdots & 1 & 0 \\ 0 & 0 & 0 & 0 & \cdots & 0 & 1 \\ -a_0 & -a_1 & -a_2 & -a_3 & \cdots & -a_{n-2} & -a_{n-1} \end{bmatrix}$$

LEMMA 19.1. *Let α be the companion matrix of the monic polynomial $f(x) \in F[x]$. Then*

$$\varphi_\alpha(x) = f(x)$$

PROOF. Let f and α be as above. To find $\varphi_\alpha(x)$ we must compute $\det \beta$ where $\beta \in F(x)^{n \times n}$ is given by

$$\beta = xI_n - \alpha = \begin{bmatrix} x & -1 & 0 & 0 & \cdots & 0 & 0 \\ 0 & x & -1 & 0 & \cdots & 0 & 0 \\ 0 & 0 & x & -1 & \cdots & 0 & 0 \\ & & & \cdots\cdots\cdots\cdots\cdots & & \\ 0 & 0 & 0 & 0 & \cdots & -1 & 0 \\ 0 & 0 & 0 & 0 & \cdots & x & -1 \\ a_0 & a_1 & a_2 & a_3 & \cdots & a_{n-2} & x + a_{n-1} \end{bmatrix}$$

We prove that $\det \beta = f(x)$ by induction on n.

Suppose $n = 1$. Then we have $f(x) = x + a_{n-1}$, $\alpha = [-a_{n-1}]$ and $\beta = [x + a_{n-1}]$, so the result follows in this case. Now suppose that $n > 1$ and the result is true for all monic polynomials of degree $n - 1$. Since there are only two nonzero entries in the first column of β, it is a simple matter to apply the cofactor expansion here and obtain

$$\det \beta = x(\det \beta_{11}) + (-1)^{n+1}a_0(\det \beta_{n1})$$

Now

$$\beta_{11} = \begin{bmatrix} x & -1 & 0 & \cdots & 0 & 0 \\ 0 & x & -1 & \cdots & 0 & 0 \\ & & & \cdots\cdots\cdots & & \\ 0 & 0 & 0 & \cdots & -1 & 0 \\ 0 & 0 & 0 & \cdots & x & -1 \\ a_1 & a_2 & a_3 & \cdots & a_{n-2} & x+a_{n-1} \end{bmatrix}$$

so $\beta_{11} = xI_{n-1} - \alpha^*$ where α^* is the companion matrix of the polynomial

$$f^*(x) = a_1 + a_2 x + \cdots + a_{n-2}x^{n-3} + a_{n-1}x^{n-2} + x^{n-1}$$

Thus by induction, $\det \beta_{11} = f^*(x)$.

On the other hand

$$\beta_{n1} = \begin{bmatrix} -1 & 0 & 0 & \cdots & 0 & 0 \\ x & -1 & 0 & \cdots & 0 & 0 \\ 0 & x & -1 & \cdots & 0 & 0 \\ & & & \cdots\cdots\cdots & & \\ 0 & 0 & 0 & \cdots & -1 & 0 \\ 0 & 0 & 0 & \cdots & x & -1 \end{bmatrix}$$

and we have seen such matrices before, where all entries above the main diagonal are zero or perhaps where all entries below the main diagonal are zero. We know that the determinant of such a matrix is equal to the product of its main diagonal entries. However here we can take a different approach. First add x times the first row to the second, then add x times the second row to the third, and continue in this manner, finally adding x times the $(n-2)$nd row to the $(n-1)$st. In this way, we have found a new matrix with the same determinant as β_{n1} and this new matrix is clearly $-I_{n-1}$. Thus we have

$$\det \beta_{n1} = \det(-I_{n-1}) = (-1)^{n-1}$$

Putting this all together, we have

$$\varphi_\alpha(x) = \det \beta = x(\det \beta_{11}) + (-1)^{n+1}a_0(\det \beta_{n1})$$
$$= x(a_1 + a_2 x + \cdots + a_{n-1}x^{n-2} + x^{n-1}) + (-1)^{n+1}a_0(-1)^{n-1}$$
$$= a_0 + a_1 x + a_2 x^2 + \cdots + a_{n-1}x^{n-1} + x^n = f(x)$$

and the lemma is proved.   □

We can conclude form the above that a field F is algebraically closed if and only if all linear transformations on finite dimensional vector spaces over F have at least one eigenvalue. Indeed, suppose first that F is algebraically closed. Then we know, by Corollary 18.2, that $T: V \to V$ has $n = \dim_F V$ eigenvalues counting multiplicities. In the other direction, let $f(x) \in F[x]$ be a nonconstant polynomial and we wish to show that $f(x)$ has a root in F. To do this, it clearly suffices to consider any nonzero scalar multiple of $f(x)$ or in other words we may assume that $f(x)$ is monic of degree n for some $n \geq 1$. Let α be its companion matrix and let $T \colon F^n \to F^n$ be a linear transformation with $\alpha = {}_{\mathcal{A}}[T]_{\mathcal{A}}$ for some basis $\mathcal{A}$ of F^n. By assumption, T has an eigenvalue in F so $\varphi_T(x) = \varphi_\alpha(x) = f(x)$ has a root in F. This implies that F is algebraically closed.

Now let us turn to a somewhat different problem. Let $f(x) = a_0 + a_1 x + \cdots + a_n x^n \in F[x]$ and let $T \in \mathcal{L}(V,V)$. Does it make sense to substitute T in for x and consider $f(T)$? The answer is "yes" with one proviso. We must consider the constant term a_0 as $a_0 x^0$ and then put $T^0 = I$, the identity linear transformation. Since $\mathcal{L}(V,V)$ is a vector space over F, we then see that

$$ f(T) = a_0 T^0 + a_1 T + a_2 T^2 + \cdots + a_n T^n $$

is just some element of $\mathcal{L}(V,V)$. We remark that of course the power T^i is of course given by

$$ T^i = \underbrace{T \cdot T \cdots T}_{i \text{ times}} $$

Furthermore, since multiplication in $\mathcal{L}(V,V)$ is associative and $T^0 = I$, we have $T^i \cdot T^j = T^{i+j}$ for all $i, j \geq 0$.

LEMMA 19.2. *Let $S, T \in \mathcal{L}(V,V)$ where V is a vector space over F.*

    i. If $a, b \in F$, then $(aS)(bT) = (ab)(ST)$.

    ii. If $h(x) = f(x)g(x)$ are polynomials in $F[x]$, then by way of evaluation at T we have $h(T) = f(T)g(T)$.

PROOF. (i) This follows immediately from Lemma 12.1. Indeed by that result, we have

$$ (aS)(bT) = a(S(bT)) = a(b(ST)) = (ab)(ST) $$

(ii) Write

$$ f(x) = \sum_i a_i x^i, \qquad g(x) = \sum_j b_j x^j $$

Since the distributive law holds in $\mathcal{L}(V,V)$, we have

$$ f(T)g(T) = \left(\sum_i a_i T^i \right) \left(\sum_j b_j T^j \right) = \sum_{i,j} (a_i T^i)(b_j T^j) $$

Now by (i) above

$$ (a_i T^i)(b_j T^j) = (a_i b_j) T^i T^j = (a_i b_j) T^{i+j} $$

so we have

$$f(T)g(T) = \sum_{i,j} (a_i b_j) T^{i+j}$$

$$= \sum_k \left(\sum_{i+j=k} a_i b_j \right) T^k = h(T)$$

and the lemma is proved.   □

Suppose $\dim_F V = n$. Then $\dim_F \mathcal{L}(V, V) = n^2$ by Theorem 11.2. Consider the $n^2 + 1$ linear transformations $T^0 = I, T^1, T^2, \ldots, T^{n^2}$. Since $n^2 + 1 > \dim_F \mathcal{L}(V, V)$, this set must be linearly dependent. Thus there exist elements $a_i \in F$ not all zero with

$$a_0 T^0 + a_1 T^1 + a_2 T^2 + \cdots + a_{n^2} T^{n^2} = 0$$

In other words, there exists a nonzero polynomial $f(x) \in F[x]$ of degree $\leq n^2$ with $f(T) = 0$.

As it turns out, the bound of n^2 on $\deg f(x)$ is really too large. Indeed, we can always find a monic polynomial $f(x)$ of degree n with $f(T) = 0$. In fact, we can take f to be the characteristic polynomial of T. This is the content of the *Cayley-Hamilton Theorem* below.

THEOREM 19.1. *Let $T\colon V \to V$ be a linear transformation with V a finite dimensional vector space over F. Then $\varphi_T(T) = 0$.*

PROOF. Suppose that $\dim_F V = n$. Since $\varphi_T(T)$ is a linear transformation, we see that $\varphi_T(T) = 0$ if and only if $\alpha \cdot \varphi_T(T) = 0$ for all $\alpha \in V$. Now $\alpha \cdot \varphi_T(T) = 0$ is certainly true for $\alpha = 0$, so we need only consider the nonzero elements of V. Let $\alpha \in V$ with $\alpha \neq 0$. Then $\{\alpha\}$ is a linearly independent subset of V. It may also be true that $\{\alpha, \alpha T\}$ is linearly independent, and then possibly $\{\alpha, \alpha T, \alpha T^2\}$. Since V is finite dimensional, we see that at some point r, $\{\alpha, \alpha T, \ldots, \alpha T^{r-1}\}$ is linearly independent, but $\{\alpha, \alpha T, \ldots, \alpha T^{r-1}, \alpha T^r\}$ is linearly dependent. From the latter dependence, there exist $a_i \in F$ not all zero with

$$a_0 \alpha + a_1 (\alpha T) + \cdots + a_{r-1}(\alpha T^{r-1}) + a_r(\alpha T^r) = 0$$

Clearly $a_r \neq 0$ since otherwise we would have a dependence relation among $\alpha, \alpha T, \ldots, \alpha T^{r-1}$ which is not the case. Thus, by multiplying through by a_r^{-1} if necessary, we may assume that $a_r = 1$ so

$$a_0 \alpha + a_1 (\alpha T) + \cdots + a_{r-1}(\alpha T^{r-1}) + \alpha T^r = 0$$

Moreover since $a_i(\alpha T^i) = \alpha(a_i T^i)$, the above becomes

$$\alpha(a_0 T^0 + a_1 T^1 + \cdots + a_{r-1} T^{r-1} + T^r) = 0$$

and hence $\alpha f(T) = 0$, where

$$f(x) = a_0 + a_1 x + \cdots + a_{r-1} x^{r-1} + x^r \in F[x]$$

Now $\{\alpha, \alpha T, \ldots, \alpha T^{r-1}\}$ is linearly independent, so we can complete this set to form a basis

$$\mathcal{A} = \{\alpha, \alpha T, \ldots, \alpha T^{r-1}, \beta_1, \beta_2, \ldots, \beta_s\}$$

for V where, of course, $r + s = n$. We consider the matrix $_\mathcal{A}[T]_\mathcal{A}$. First, for $0 \le i \le r - 2$ we have $(\alpha T^i)T = \alpha T^{i+1}$ so

$$(\alpha T^i)T = 0\alpha + 0(\alpha T) + \cdots + 1(\alpha T^{i+1}) + \cdots + 0(\alpha T^{r-1})$$
$$+ 0\beta_1 + 0\beta_2 + \cdots + 0\beta_s$$

and for $i = r - 1$ we have $(\alpha T^{r-1})T = \alpha T^r$ so

$$(\alpha T^{r-1})T = -a_0\alpha - a_1(\alpha T) - \cdots - a_{r-1}(\alpha T^{r-1})$$
$$+ 0\beta_1 + 0\beta_2 + \cdots + 0\beta_s$$

Therefore in block matrix form

$$_\mathcal{A}[T]_\mathcal{A} = \left[\begin{array}{c|c} A & 0 \\ \hline C & B \end{array}\right]$$

where $B \in F^{s \times s}$, $C \in F^{s \times r}$ and

$$A = \begin{bmatrix} 0 & 1 & 0 & 0 & \cdots & 0 & 0 \\ 0 & 0 & 1 & 0 & \cdots & 0 & 0 \\ 0 & 0 & 0 & 1 & \cdots & 0 & 0 \\ & & & \cdots\cdots\cdots\cdots\cdots & & \\ 0 & 0 & 0 & 0 & \cdots & 1 & 0 \\ 0 & 0 & 0 & 0 & \cdots & 0 & 1 \\ -a_0 & -a_1 & -a_2 & -a_3 & \cdots & -a_{r-2} & -a_{r-1} \end{bmatrix}$$

is the companion matrix of the polynomial $f(x)$.

We compute $\varphi_T(x)$. By definition

$$\varphi_T(x) = \det(xI_n - {}_\mathcal{A}[T]_\mathcal{A})$$

$$= \det \left[\begin{array}{c|c} xI_r - A & 0 \\ \hline -C & xI_s - B \end{array}\right]$$

and thus by Theorem 16.3 we have

$$\varphi_T(x) = \det(xI_r - A)\cdot \det(xI_s - B) = \varphi_A(x)\varphi_B(x)$$

But A is the companion matrix of $f(x)$, so Lemma 19.1 yields $\varphi_A(x) = f(x)$ and hence $\varphi_T(x) = f(x)\varphi_B(x)$.

Finally, by part (ii) of Lemma 19.2, $\varphi_T(T) = f(T)\varphi_B(T)$, so

$$\alpha \cdot \varphi_T(T) = \alpha \cdot \big(f(T)\varphi_B(T)\big) = \big(\alpha \cdot f(T)\big)\varphi_B(T)$$
$$= 0 \cdot \varphi_B(T) = 0$$

Thus $\alpha \cdot \varphi_T(T) = 0$ for all $\alpha \in V$ and we conclude that $\varphi_T(T) = 0$.   □

In the same way, if $\gamma \in F^{n \times n}$ and $f(x) \in F[x]$, then we can substitute γ in for x with the proviso that $\gamma^0 = I_n$. Clearly $f(\gamma) \in F^{n \times n}$.

COROLLARY 19.1. *Let $\gamma \in F^{n \times n}$. Then $\varphi_\gamma(\gamma) = 0$.*

PROOF. Let $\mathcal{A}$ be a basis for F^n and define a linear transformation $T \colon F^n \to F^n$ by $_{\mathcal{A}}[T]_{\mathcal{A}} = \gamma$. Since

$$\gamma^i = \gamma \cdot \gamma \cdots \gamma = {}_{\mathcal{A}}[T]_{\mathcal{A}} \cdot {}_{\mathcal{A}}[T]_{\mathcal{A}} \cdots {}_{\mathcal{A}}[T]_{\mathcal{A}} = {}_{\mathcal{A}}[T^i]_{\mathcal{A}}$$

we have clearly $f(\gamma) = {}_{\mathcal{A}}[f(T)]_{\mathcal{A}}$ for any polynomial $f(x)$. Finally, $\varphi_T(x) = \varphi_\gamma(x)$ so, by the Cayley-Hamilton Theorem,

$$\varphi_\gamma(\gamma) = \varphi_T(\gamma) = {}_{\mathcal{A}}[\varphi_T(T)]_{\mathcal{A}} = {}_{\mathcal{A}}[0]_{\mathcal{A}} = 0$$

and the result follows.   □

Problems

A nonempty subset A of $F[x]$ is said to be an *ideal* of the ring if

   i. $\alpha, \beta \in A$ implies that $\alpha + \beta \in A$.

  ii. $\alpha \in A$ and $\beta \in F[x]$ implies that $\alpha\beta \in A$.

For example $A = \{0\}$ is an ideal and so is the set of all $F[x]$-multiples of any fixed element of the ring. In the following three problems, let A be an ideal of $F[x]$ with $A \neq \{0\}$.

19.1. Let m be the minimal degree of all nonzero polynomials in A. Show that A contains a unique monic polynomial $\mu(x)$ of degree m. μ is called the *minimal polynomial* of A.

19.2. Show that for all $f(x) \in A$, there exists some polynomial $g(x) \in F[x]$ with $f(x) = \mu(x)g(x)$. (Hint, apply induction on the degree of $f(x)$. If $\deg f = n \geq m$ and $f(x) = a_n x^n + $ lower degree terms, then $f(x) - a_n x^{n-m}\mu(x) \in A$ has degree less than n.)

19.3. Conclude that $A = F[x]\mu(x)$ is the set of all $F[x]$-multiples of its minimal polynomial $\mu(x)$.

Let V be a finite dimensional vector space over F and let $T: V \to V$ be a linear transformation. Define

$$A_T = \{f(x) \in F[x] \mid f(T) = 0\}$$

19.4. Show that A_T is an ideal of $F[x]$ with $A_T \neq \{0\}$. We call the minimal polynomial $\mu_T(x)$ of A_T the *minimal polynomial* of T. Prove that $\mu_T(x)$ divides the characteristic polynomial $\varphi_T(x)$.

19.5. Let $\mathcal{A}$ be a basis for V and let $\alpha = {}_{\mathcal{A}}[T]_{\mathcal{A}}$. Prove that $f(x) \in A_T$ if and only if $f(\alpha) = 0$.

19.6. Let $f(x), g(x) \in F[x]$.
   i. If $f(x)$ and $g(x)$ have no roots in F, show that the same is true for $f(x)g(x)$.
   ii. If $f(x)$ is monic and divides $\prod_1^n (x - a_i)$, prove that $f(x)$ is a partial product of the $(x - a_i)$ factors.

Using the fact that $\mu_T(x)$ divides $\varphi_T(x)$, find the minimal polynomials of the linear transformations below given by their corresponding matrices.

19.7.

$$_{\mathcal{A}}[T]_{\mathcal{A}} = \begin{bmatrix} 1 & 0 & 0 & 0 \\ 0 & 1 & 0 & 0 \\ 0 & 0 & 1 & 0 \\ 0 & 0 & 0 & 2 \end{bmatrix} \in \mathbb{R}^{4 \times 4}$$

19.8.

$$_{\mathcal{A}}[T]_{\mathcal{A}} = \mathrm{diag}(a_1, a_2, \ldots, a_n)$$

with $a_1, a_2, \ldots, a_n$ distinct elements of F.

19.9.

$$_{\mathcal{A}}[T]_{\mathcal{A}} = \begin{bmatrix} 0 & 0 & 0 & 0 \\ 1 & 0 & 0 & 0 \\ 0 & 1 & 0 & 0 \\ 0 & 0 & 1 & 0 \end{bmatrix} \in F^{4 \times 4}$$

19.10.

$$_{\mathcal{A}}[T]_{\mathcal{A}} = \begin{bmatrix} 2 & 0 & 0 \\ 1 & 2 & 0 \\ 0 & 1 & 2 \end{bmatrix} \in \mathbb{R}^{3 \times 3}$$

20. Nilpotent Transformations

Let $T\colon V \to V$ be a linear transformation and suppose that its characteristic polynomial is given by

$$\varphi_T(x) = (x - a_1)(x - a_2)\cdots(x - a_n)$$

If $a_1, a_2, \ldots, a_n$ are all distinct elements of F, then as we have seen, T can be diagonalized. What can we conclude if all the roots are not distinct? Obviously the worst offenders are those T that satisfy $a_1 = a_2 = \cdots = a_n$ and we consider the case now where this common value is 0. Thus $\varphi_T(x) = x^n$ and, by the Cayley-Hamilton Theorem, we have $T^n = \varphi_T(T) = 0$. In other words, T is a *nilpotent transformation*. Formally, a nilpotent transformation is a linear transformation $T\colon V \to V$ satisfying $T^m = 0$ for some integer m. Let us consider some examples.

EXAMPLE 20.1. Let V denote the subspace of $\mathbb{R}[x]$ consisting of all polynomials of degree $< n$, for some fixed n, and let $D\colon V \to V$ be the derivative map. Since applying D lowers the degree of a nonzero polynomial and since $\dim_{\mathbb{R}} V = n$, we see that $D^n = 0$. Now V has a basis

$$\mathcal{A} = \left\{ \frac{x^{n-1}}{(n-1)!}, \frac{x^{n-2}}{(n-2)!}, \ldots, \frac{x^2}{2!}, x, 1 \right\}$$

and since $(x^i/i!)D = x^{i-1}/(i-1)!$ for $i \geq 1$, and $1D = 0$, we see that

$$_{\mathcal{A}}[D]_{\mathcal{A}} = \begin{bmatrix} 0 & 1 & 0 & 0 & \cdots & 0 \\ 0 & 0 & 1 & 0 & \cdots & 0 \\ 0 & 0 & 0 & 1 & \cdots & 0 \\ & & \cdots\cdots\cdots\cdots & & \\ 0 & 0 & 0 & 0 & \cdots & 1 \\ 0 & 0 & 0 & 0 & \cdots & 0 \end{bmatrix}$$

EXAMPLE 20.2. Again let V be as above and define $\Delta\colon V \to V$ by

$$\alpha(x)\Delta = \alpha(x + 1) - \alpha(x)$$

Then Δ is called a *difference operator* and it is in some sense a finite analog of the derivative. Observe that for $i > 0$

$$x^i \Delta = (x + 1)^i - x^i$$

has degree $i - 1$, so applying Δ always lowers the degree of a nonzero polynomial. This therefore implies that $\Delta^n = 0$. Now for $i \geq 0$, define $x^{(i)}$ by $x^{(i)} = x(x-1)\cdots(x-i+1)$ with $x^{(1)} = x$ and $x^{(0)} = 1$. Then for $i > 0$, we have

$$x^{(i)}\Delta = (x+1)x(x-1)\cdots(x-i+2) - x(x-1)\cdots(x-i+1)$$
$$= \Big((x+1) - (x-i+1)\Big)x(x-1)\cdots(x-i+2)$$
$$= ix^{(i-1)}$$

Since

$$\mathcal{B} = \left\{ \frac{x^{(n-1)}}{(n-1)!}, \frac{x^{(n-2)}}{(n-2)!}, \ldots, \frac{x^{(2)}}{2!}, x^{(1)}, x^{(0)} \right\}$$

is also a basis for V, we see easily that

$$_\mathcal{B}[\Delta]_\mathcal{B} = \begin{bmatrix} 0 & 1 & 0 & 0 & \cdots & 0 \\ 0 & 0 & 1 & 0 & \cdots & 0 \\ 0 & 0 & 0 & 1 & \cdots & 0 \\ & & & \cdots & & \\ 0 & 0 & 0 & 0 & \cdots & 1 \\ 0 & 0 & 0 & 0 & \cdots & 0 \end{bmatrix}$$

EXAMPLE 20.3. Let us return to the formal derivative map, but this time over a different field. Indeed, let F be the field of two elements. Thus $F = \{0, 1\}$ and $1 + 1 = 0$. If $\alpha(x) = \sum_{i\geq 0} a_i x^i \in F[x]$, then $\alpha(x)D = \sum_{i\geq 1} i a_i x^{i-1}$ and

$$\alpha(x)D^2 = \sum_{i\geq 2} i(i-1)a_i x^{i-2}$$

Now $i(i-1)$ is an even integer, so $i(i-1)$ is a sum of an even number of identity elements of F. Since $1 + 1 = 0$, we therefore have $i(i-1) = 0$ in F and hence $\alpha(x)D^2 = 0$. Thus $D^2 = 0$. Now suppose V is the subspace of $F[x]$ consisting of all polynomials of degree ≤ 4. Then $\mathcal{C} = \{x^4, x^3, x^2, x, 1\}$ is a basis for V and

$$_\mathcal{C}[D]_\mathcal{C} = \begin{bmatrix} 0 & 0 & 0 & 0 & 0 \\ 0 & 0 & 1 & 0 & 0 \\ 0 & 0 & 0 & 0 & 0 \\ 0 & 0 & 0 & 0 & 1 \\ 0 & 0 & 0 & 0 & 0 \end{bmatrix}$$

Thus we have seen in all of the above examples that a basis $\mathcal{B}$ exists for which $_\mathcal{B}[T]_\mathcal{B}$ is a matrix with zero entries everywhere but on the diagonal directly above the main diagonal. Moreover on that *super diagonal* the entries are either 0 or 1. We will see in the following that this is always the case for nilpotent transformations.

Let $a \in F$ and let $s \geq 1$ be an integer. Then we call the $s \times s$ F-matrix

$$\begin{bmatrix} a & 1 & 0 & 0 & \cdots & 0 \\ 0 & a & 1 & 0 & \cdots & 0 \\ 0 & 0 & a & 1 & \cdots & 0 \\ & & & \cdots & & \\ 0 & 0 & 0 & 0 & \cdots & 1 \\ 0 & 0 & 0 & 0 & \cdots & a \end{bmatrix}$$

an (a, s)-*Jordan block* and we denote it by $\mathfrak{b}(a, s)$. Specifically, $\mathfrak{b}(a, s)$ has all main diagonal entries equal to a, all super diagonal entries equal to 1, and zeros elsewhere. Of course, if $s = 1$, then $\mathfrak{b}(a, 1) = [a]$. Next, we denote by

$$\mathrm{diag}(\mathfrak{b}(a_1, s_1), \mathfrak{b}(a_2, s_2), \ldots, \mathfrak{b}(a_r, s_r))$$

the $n \times n$ block diagonal matrix given pictorially by

$$\begin{bmatrix} \mathfrak{b}(a_1, s_1) & & & 0 \\ & \mathfrak{b}(a_2, s_2) & & \\ & & \ddots & \\ 0 & & & \mathfrak{b}(a_r, s_r) \end{bmatrix}$$

where $n = s_1 + s_2 + \cdots + s_r$. The main result of this section is

THEOREM 20.1. *Let V be a finite dimensional vector space over F and let $T: V \to V$ be a nilpotent linear transformation. Then there exists a basis $\mathcal{B}$ of V such that*

$$_\mathcal{B}[T]_\mathcal{B} = \mathrm{diag}(\mathfrak{b}(0, s_1), \mathfrak{b}(0, s_2), \ldots, \mathfrak{b}(0, s_r))$$

with $s_1 \geq s_2 \geq \cdots \geq s_r$.

The proof of this theorem is elementary, but somewhat tedious. If $\mathcal{B}$ is a basis for V with $_\mathcal{B}[T]_\mathcal{B}$ having the appropriate form, then it is easy to see that $\mathcal{B} \cap \ker T$ is a basis for $\ker T$, and that the size of $\mathcal{B} \cap \ker T$ is equal to r, the number of blocks in the block diagonal matrix. Thus it is reasonable to suppose that the proof should start with a basis for $\ker T$ and then use each basis member to construct a corresponding "block" of vectors in $\mathcal{B}$. As will be apparent, we have to be a bit careful in the choice of this starting basis for $\ker T$.

PROOF. We first define a descending chain of subspaces of V as follows. Let $V_i = \mathrm{im}\, T^i = (V)T^i$ so that V_i is a subspace of V and $V_0 = V$. Moreover for $i > 0$

$$V_i = (V)T^i = (VT)T^{i-1} \subseteq (V)T^{i-1} = V_{i-1}$$

so $V_i \subseteq V_{i-1}$. Finally, T is nilpotent, so $T^{n+1} = 0$ for some $n + 1 \geq 1$ and hence $V_{n+1} = 0$. Thus we have

$$V = V_0 \supseteq V_1 \supseteq V_2 \supseteq \cdots \supseteq V_n \supseteq V_{n+1} = 0$$

Next, let $W = \ker T$ and set $W_i = W \cap V_i$. Then clearly

$$W = W_0 \supseteq W_1 \supseteq W_2 \supseteq \cdots \supseteq W_n \supseteq W_{n+1} = 0$$

At this point, we choose a basis for W consistent with this chain of subspaces. Specifically, we start with a basis for W_n, extend it to a basis for W_{n-1}, extend that to a basis for W_{n-2} and continue this process. In this way, we obtain a basis

$$\{\alpha_{ij} \mid i = 0, 1, \ldots, n; \ j = 0, 1, \ldots, f(i)\}$$

for W such that $\alpha_{ij} \in W_i \setminus W_{i+1}$ and such that, for each integer q, $\{\alpha_{ij} \mid i \geq q\}$ is a basis for W_q.

Since $\alpha_{ij} \in V_i = \operatorname{im} T^i$, we can choose $\beta_{ij} \in V$ with $\beta_{ij}T^i = \alpha_{ij}$. For each such i, j we then obtain $i + 1$ potentially different vectors of V, namely $\beta_{ij}, \beta_{ij}T, \beta_{ij}T^2, \ldots, \beta_{ij}T^i = \alpha_{ij}$. We show now that the set of vectors

$$\mathcal{B} = \{\beta_{ij}T^k \mid i = 0, 1, \ldots, n; \ j = 0, 1, \ldots, f(i); \ k = 0, 1, \ldots, i\}$$

is a basis for V.

We first prove that the vectors in $\mathcal{B}$ are distinct and linearly independent. Thus suppose

$$\sum_{i,j,k} a_{ijk}(\beta_{ij}T^k) = 0$$

with $a_{ijk} \in F$. If some a_{ijk} is not zero, let

$$m = \max\{i - k \mid a_{ijk} \neq 0\}$$

so that $m \geq 0$ since we always have $i \geq k$. In particular, if $i - k > m$ then $a_{ijk} = 0$, but for some i, j, k with $i - k = m$ we have $a_{ijk} \neq 0$. We now apply T^m to the given linear dependence. Since T^m is a linear transformation, we obtain

$$\sum_{i,j,k} a_{ijk}(\beta_{ij}T^{k+m}) = \left(\sum_{i,j,k} a_{ijk}(\beta_{ij}T^k)\right)T^m = 0T^m = 0$$

Let us consider each term $a_{ijk}(\beta_{ij}T^{k+m})$ in this sum. If $i - k > m$ then, by assumption, $a_{ijk} = 0$ so this term is 0. On the other hand, if $i - k < m$, then $i < k + m$ so

$$\beta_{ij}T^{k+m} = (\beta_{ij}T^i)T \cdot T^{k+m-i-1}$$

$$= (\alpha_{ij}T) \cdot T^{k+m-i-1} = 0T^{k+m-i-1} = 0$$

since $\beta_{ij}T^i = \alpha_{ij} \in W = \ker T$. Thus the only terms that occur in the above sum have $i - k = m$, so $i = k + m$ and

$$a_{ijk}(\beta_{ij}T^{k+m}) = a_{ijk}(\beta_{ij}T^i) = a_{ijk}\alpha_{ij}$$

Thus we have

$$\sum_{\substack{i,j,k \\ k=i-m}} a_{ijk}\alpha_{ij} = 0$$

But $\{\alpha_{ij}\}$ is a basis for W and there is at most one subscript k for each i, j, so we conclude that all $a_{ijk} = 0$ with $i - k = m$, a contradiction by the definition of m. This proves that $\mathcal{B}$ is a linearly independent set of distinct vectors.

We show now, by inverse induction on $q = n + 1, n, \ldots, 0$, that the subset of $\mathcal{B}$ given by $\mathcal{B}_q = \{\beta_{ij}T^k \mid k \geq q\}$ spans $V_q = \operatorname{im} T^q = (V)T^q$. Of course, $\mathcal{B}_q \subseteq V_q$ since all the exponents of T in the vectors of $\mathcal{B}_q$ are at least equal to q. If $q = n+1$, then $V_q = 0$ and $\mathcal{B}_q = \emptyset$, so the result is trivially true. Let us now suppose that $\mathcal{B}_{q+1}$ spans V_{q+1} and let $\gamma \in V_q$. Since $\gamma \in V_q = \operatorname{im} T^q$, it follows easily that $\gamma T \in \operatorname{im} T^{q+1} = V_{q+1}$ and thus, by induction,

$$\gamma T = \sum_{\substack{i,j,k \\ k \geq q+1}} b_{ijk}(\beta_{ij}T^k)$$

for suitable $b_{ijk} \in F$. Note that $q \geq 0$ implies that $k \geq q + 1 \geq 1$ and thus if we define

$$\delta = \gamma - \sum_{\substack{i,j,k \\ k \geq q+1}} b_{ijk}(\beta_{ij}T^{k-1})$$

then $\delta T = 0$.

Two facts about $\delta \in V$ are now apparent. First, $\delta T = 0$ so $\delta \in \ker T = W$. Second, $\delta \in V_q$ since $\gamma \in V_q$ and since $k-1 \geq q$ implies that $\beta_{ij}T^{k-1} \in V_q$. Thus $\delta \in W \cap V_q = W_q$. Now we know that $\{\alpha_{ij} \mid i \geq q\}$ is a basis for W_q and therefore, for suitable elements $c_{ij} \in F$, we have

$$\delta = \sum_{\substack{i,j \\ i \geq q}} c_{ij}\alpha_{ij} = \sum_{\substack{i,j \\ i \geq q}} c_{ij}(\beta_{ij}T^i)$$

Since

$$\gamma = \delta + \sum_{\substack{i,j,k \\ k \geq q+1}} b_{ijk}(\beta_{ij}T^{k-1})$$

$$= \sum_{\substack{i,j \\ i \geq q}} c_{ij}(\beta_{ij}T^i) + \sum_{\substack{i,j,k \\ k \geq q+1}} b_{ijk}(\beta_{ij}T^{k-1})$$

and since all exponents of T are at least q in size, we conclude that $\mathcal{B}_q$ spans V_q and the induction step is proved. Finally, $V = V_0$, so $\mathcal{B} = \mathcal{B}_0$ spans V and therefore $\mathcal{B}$ is a basis for V.

We finish the proof by writing T in terms of the basis $\mathcal{B}$. However, this requires that we first choose an ordering for the members of the basis. We do this in such a way that $\beta_{ij}, \beta_{ij}T, \beta_{ij}T^2, \cdots, \beta_{ij}T^i$ occur consecutively. We then have to decide in which order these β_{ij} "blocks" occur and we do this in such a way that the larger i subscripts appear ahead of the smaller ones. It is now a simple matter to write down the matrix $_\mathcal{B}[T]_\mathcal{B}$. Observe that for $k < i$ we have $(\beta_{ij}T^k)T = \beta_{ij}T^{k+1}$, and that for $k = i$ we have $(\beta_{ij}T^k)T = \alpha_{ij}T = 0$ since $\alpha_{ij} \in W = \ker T$. Thus the contribution of this "block" of the basis to the matrix $_\mathcal{B}[T]_\mathcal{B}$ of T is the $(i+1) \times (i+1)$ matrix

$$
\mathfrak{b}(0, i+1) = \begin{bmatrix} 0 & 1 & 0 & 0 & \cdots & 0 \\ 0 & 0 & 1 & 0 & \cdots & 0 \\ 0 & 0 & 0 & 1 & \cdots & 0 \\ & & \cdots\cdots\cdots & & \\ 0 & 0 & 0 & 0 & \cdots & 1 \\ 0 & 0 & 0 & 0 & \cdots & 0 \end{bmatrix}
$$

and this occurs as a block on the main diagonal.

Thus we see finally that

$$
_\mathcal{B}[T]_\mathcal{B} = \mathrm{diag}(\mathfrak{b}(0, s_1), \mathfrak{b}(0, s_2), \ldots, \mathfrak{b}(0, s_r))
$$

where each Jordan block $\mathfrak{b}(0, s_t)$ corresponds to some "block"

$$
\beta_{ij}, \beta_{ij}T, \beta_{ij}T^2, \cdots, \beta_{ij}T^i
$$

of the basis. Therefore $s_t = i + 1$ and, since we have ordered the β_{ij}s in such a way that the larger i subscripts appear first, we see that $s_1 \geq s_2 \geq \cdots \geq s_r$. The theorem is proved. $\qquad\square$

It is interesting to observe that the integers $s_1, s_2, \ldots, s_r$ are actually uniquely determined by T. A proof of this fact is given in the exercises at the end of this section.

COROLLARY 20.1. *Let V be a finite dimensional vector space over F and let $T \colon V \to V$ be a linear transformation. Let $a \in F$ and suppose that $T - aI$ is nilpotent. Then there exists a basis $\mathcal{B}$ of V such that*

$$
_\mathcal{B}[T]_\mathcal{B} = \mathrm{diag}(\mathfrak{b}(a, s_1), \mathfrak{b}(a, s_2), \ldots, \mathfrak{b}(a, s_r))
$$

with $s_1 \geq s_2 \geq \cdots \geq s_r$.

PROOF. If $S = T - aI$, then by the preceding theorem we can find a basis $\mathcal{B}$ with

$$\mathcal{B}[S]_\mathcal{B} = \text{diag}(\mathfrak{b}(0, s_1), \mathfrak{b}(0, s_2), \ldots, \mathfrak{b}(0, s_r))$$

and with $s_1 \geq s_2 \geq \cdots \geq s_r$. Now $T = aI + S$ so

$$\mathcal{B}[T]_\mathcal{B} = \mathcal{B}[aI]_\mathcal{B} + \mathcal{B}[S]_\mathcal{B} = aI_n + \mathcal{B}[S]_\mathcal{B}$$

where $n = \dim_F V$. Thus, since clearly $aI_s + \mathfrak{b}(0, s) = \mathfrak{b}(a, s)$, the result follows. $\qquad\square$

Problems

20.1. Let F be a field with $1 + 1 + 1 = 0$. Discuss the formal derivative map and the difference map on the full polynomial ring $F[x]$ and show that these maps are nilpotent.

Recall that R is a *commutative ring* if it is a set with an addition and multiplication that satisfies all the axioms for a field with the possible exception of the existence of multiplicative inverses.

20.2. If $p \in \mathbb{Z}$ is a prime, show that p divides the binomial coefficients $\binom{p}{i}$ for $i = 1, 2, \ldots, p - 1$. Now let R be a commutative ring with

$$p = \underbrace{1 + 1 + \cdots + 1}_{p \text{ times}} = 0$$

Deduce that $(a + b)^p = a^p + b^p$ for all $a, b \in R$.

20.3. Let V be an F-vector space and let $T\colon V \to V$ be a nilpotent linear transformation. Let the integer $m \geq 1$ be minimal with $T^m = 0$. If $f(x) \in F[x]$ with $f(T) = 0$, show that $f(x) = x^m g(x)$ for some polynomial $g(x)$. See Problems 19.4 and 19.6.

Now let V be a finite dimensional vector space over F with basis $\mathcal{B}$ and let $T\colon V \to V$ be a linear transformation.

20.4. If $\mathcal{B}[T]_\mathcal{B} = \mathfrak{b}(0, n)$ show that

$$\text{rank } T^k = \begin{cases} n - k, & \text{for } k \leq n \\ 0, & \text{for } k > n \end{cases}$$

20.5. Now suppose that

$$\mathcal{B}[T]_\mathcal{B} = \text{diag}(\mathfrak{b}(0, s_1), \mathfrak{b}(0, s_2), \ldots, \mathfrak{b}(0, s_r))$$

and let n_s denote the number of block sizes s_j that are equal to s. Use the preceding result to show that

$$\text{rank } T^k = \sum_{s \geq k} n_s(s-k) = n_{k+1} + 2n_{k+2} + 3n_{k+3} + \cdots$$

and deduce that

$$\text{rank } T^k - \text{rank } T^{k+1} = n_{k+1} + n_{k+2} + n_{k+3} + \cdots$$

As a check, we observe that

$$r = n_1 + n_2 + n_3 + \cdots = \text{rank } T^0 - \text{rank } T^1$$
$$= \dim_F V - \dim_F (V)T = \dim_F \ker T$$

20.6. Deduce from the above that for $s \geq 1$

$$n_s = (\text{rank } T^{s-1} - \text{rank } T^s) - (\text{rank } T^s - \text{rank } T^{s+1})$$
$$= \text{rank } T^{s-1} - 2 \text{ rank } T^s + \text{rank } T^{s+1}$$

and therefore conclude that the integers $s_1, s_2, \ldots, s_r$ are uniquely determined by T.

Let $T \colon V \to V$. If $V = U \oplus W$ is the direct sum of two nonzero subspaces with $(U)T \subseteq U$ and $(W)T \subseteq W$, then we say that V is *decomposable* with respect to T. Otherwise it is *indecomposable*.

20.7. Suppose V is finite dimensional and decomposable as above. Let $\mathcal{A}$ be a basis for U and let $\mathcal{C}$ be a basis for W. If $\mathcal{B} = \mathcal{A} \cup \mathcal{C}$, describe the matrix $_\mathcal{B}[T]_\mathcal{B}$. What happens when V is a finite direct sum of T-stable subspaces?

20.8. Let V be finite dimensional and assume that T is nilpotent. Show that V is indecomposable with respect to T if and only if there exists a basis $\mathcal{B}$ with $_\mathcal{B}[T]_\mathcal{B} = \mathfrak{b}(0, n)$. Hint. If $_\mathcal{B}[T]_\mathcal{B} = \mathfrak{b}(0, n)$, first observe that $\ker T$ is 1-dimensional.

Let $S, T \colon V \to V$ be commuting linear transformations so that $ST = TS$.

20.9. If $W = (V)S^k$ or if $W = (0)\overleftarrow{S^k}$, show that $(W)T \subseteq W$.

20.10. Suppose S and T are both nilpotent. Prove that $S + T$ is also a nilpotent transformation. What happens if S and T do not commute?

21. Jordan Canonical Form

Let $T\colon V \to V$ be a linear transformation with characteristic polynomial given by $\varphi_T(x) = (x - a_1)(x - a_2)\cdots(x - a_n)$. If all the eigenvalues a_i are distinct then, as we have seen previously, T can be diagonalized. If all the roots a_i are identical, then the results of the last section imply that T can be diagonalized by Jordan block matrices. These assumptions are obviously the extreme cases and in this section we finally prove a general result. The proof will be by induction on $\dim_F V$ and therefore we will have to consider T acting on subspaces of V.

Suppose $T\colon V \to V$ and let W be a subspace of V. We say that W is *T-invariant* or *T-stable* if $(W)T \subseteq W$. In this case, by restricting our attention to W, we can clearly view $T\colon W \to W$ as a linear transformation. To avoid confusion, we denote this restricted transformation by T_W. Let us consider some simple examples.

EXAMPLE 21.1. Suppose α is an eigenvector of T with corresponding eigenvalue a. If $W = \langle \alpha \rangle$, then W is a 1-dimensional T-invariant subspace of V. It is easy to see that $T_W\colon W \to W$ is just scalar multiplication by $a \in F$.

EXAMPLE 21.2. Let $W = \operatorname{im} T = (V)T$. Then $(W)T = ((V)T)T \subseteq (V)T$, so W is T-invariant. In addition if $W' = \ker T = (0)T$, then $(W')T = 0 \subseteq W'$, so W' is also T-stable. Clearly $T_{W'} = 0$.

We say that V is *decomposable* with respect to T if $V = U \oplus W$ is a direct sum of two nonzero T-invariant subspaces U and W. If no such decomposition exists, then V is *indecomposable*. If V is decomposable, it is a simple matter to describe T in terms of its action on the T-stable subspaces.

LEMMA 21.1. *Let $T\colon V \to V$ be a linear transformation on a finite dimensional vector space V and suppose that $V = U \oplus W$ is a direct sum of two nonzero T-invariant subspaces. Let $\mathcal{A}$ be a basis for U and let $\mathcal{C}$ be a basis for W. Then $\mathcal{B} = \mathcal{A} \cup \mathcal{C}$ is a basis for V and we have pictorially*

$$
{}_\mathcal{B}[T]_\mathcal{B} = \left[\begin{array}{c|c} {}_\mathcal{A}[T_U]_\mathcal{A} & 0 \\ \hline 0 & {}_\mathcal{C}[T_W]_\mathcal{C} \end{array}\right]
$$

PROOF. Let $\mathcal{A} = \{\alpha_1, \alpha_2, \ldots, \alpha_r\}$ and $\mathcal{C} = \{\gamma_1, \gamma_2, \ldots, \gamma_s\}$. Then certainly

$$\mathcal{B} = \mathcal{A} \cup \mathcal{C} = \{\alpha_1, \alpha_2, \ldots, \alpha_r, \gamma_1, \gamma_2, \ldots, \gamma_s\}$$

is a basis for $V = U \oplus W$. Let ${}_{\mathcal{A}}[T_U]_{\mathcal{A}} = [a_{ij}]$ and ${}_{\mathcal{C}}[T_W]_{\mathcal{C}} = [c_{ij}]$. We compute ${}_{\mathcal{B}}[T]_{\mathcal{B}}$. To start with, $\alpha_i \in U$ so

$$\alpha_i T = \alpha_i T_U = \sum_j a_{ij}\alpha_j = \sum_j a_{ij}\alpha_j + \sum_k 0\gamma_k$$

Also, $\gamma_i \in W$ so

$$\gamma_i T = \gamma_i T_W = \sum_j c_{ij}\gamma_j = \sum_k 0\alpha_k + \sum_j c_{ij}\gamma_j$$

This shows that

$$
{}_{\mathcal{B}}[T]_{\mathcal{B}} =
\left[
\begin{array}{c|c}
[a_{ij}] & 0 \\
\hline
0 & [c_{ij}]
\end{array}
\right]
$$

and the lemma is proved.   □

This result will obviously be of use to us provided that we can handle the indecomposable situation. The key to this is the following.

LEMMA 21.2. *Let $T\colon V \to V$ and suppose that $\dim_F V = n < \infty$. Set $U = VT^n$ and $W = (0)\overleftarrow{T^n}$. Then U and W are both T-invariant subspaces of V, T_U is nonsingular, T_W is nilpotent, and $V = U \oplus W$.*

PROOF. Let $W = (0)\overleftarrow{T^n}$. Then

$$(WT)T^n = (WT^n)T = 0T = 0$$

so $WT \subseteq (0)\overleftarrow{T^n} = W$ and W is T-invariant. Since $WT^n = 0$, we conclude that T_W is nilpotent.

Now consider the subspaces of V given by $U_i = VT^i$ for $i = 0, 1, 2, \ldots$. Since

$$U_{i+1} = VT^{i+1} = (VT)T^i \subseteq VT^i = U_i$$

we have $V = U_0 \supseteq U_1 \supseteq U_2 \supseteq \ldots$. If U_{i+1} is properly contained in U_i, then $\dim_F U_{i+1} < \dim_F U_i$. Since $\dim_F V = n$, it follows that for some integer $k \le n$ we must have $U_{k+1} = U_k$. This then implies that

$$U_{k+2} = U_{k+1}T = U_k T = U_{k+1} = U_k$$

and continuing in this manner we see that for all $j \ge k$ we have $U_j = U_k$. In particular, since $k \le n$ we therefore have

$$U_n T = U_{n+1} = U_n$$

Let $U = U_n$. Then from the above $UT = U$ so U is T-invariant and furthermore T_U maps U onto U. Since U is finite dimensional, we conclude that T_U is nonsingular.

It remains to show finally that $V = U \oplus W$. First let $\alpha \in U \cap W$. Since $\alpha \in W$, we have $\alpha T^n = 0$. But $\alpha \in U$ and T_U is nonsingular, so we have $\alpha = 0$. Therefore $U \cap W = 0$ and $U + W = U \oplus W$. Next observe that since $T_U \colon U \to U$ is onto, so is $(T_U)^n \colon U \to U$. Let $\beta \in V$. Then $\beta T^n \in U$ and by the above remark, there exists $\alpha \in U$ with $\beta T^n = \alpha T^n$. Then $(\beta - \alpha)T^n = 0$, so $\beta - \alpha = \gamma \in W$ and $\beta = \alpha + \gamma \in U + W$. Thus

$$V = U + W = U \oplus W$$

and the lemma is proved. $\square$

We now come to the main result of this section. We obtain a description, the *Jordan canonical form*, of all linear transformations on a finite dimensional vector space over an algebraically closed field.

THEOREM 21.1. *Let V be a finite dimensional vector space over an algebraically closed field F and let $T \colon V \to V$ be a linear transformation. Then there exists a basis $\mathcal{B}$ of V such that*

$$\mathcal{B}[T]_{\mathcal{B}} = \mathrm{diag}(\mathfrak{b}(a_1, s_1), \mathfrak{b}(a_2, s_2), \ldots, \mathfrak{b}(a_r, s_r))$$

for suitable integers $s_i \geq 1$ and elements $a_i \in F$.

PROOF. We proceed by induction on $n = \dim_F V$, the result being clear for $n = 1$. Suppose first that V is decomposable with respect to T and say $V = U \oplus W$ with each of U and W a proper T-invariant subspace of V. Since $\dim_F U < n$ and $\dim_F W < n$, there exist, by induction, bases $\mathcal{A}$ of U and $\mathcal{C}$ of W with

$$\mathcal{A}[T_U]_{\mathcal{A}} = \mathrm{diag}(\mathfrak{b}(a_1, s_1), \mathfrak{b}(a_2, s_2), \ldots, \mathfrak{b}(a_t, s_t))$$
$$\mathcal{C}[T_W]_{\mathcal{C}} = \mathrm{diag}(\mathfrak{b}(a_{t+1}, s_{t+1}), \mathfrak{b}(a_{t+2}, s_{t+2}), \ldots, \mathfrak{b}(a_r, s_r))$$

for suitable integers $s_i \geq 1$ and elements $a_i \in F$. If $\mathcal{B} = \mathcal{A} \cup \mathcal{C}$ then by Lemma 21.1, $\mathcal{B}$ is a basis for V and

$$\mathcal{B}[T]_{\mathcal{B}} = \left[\begin{array}{c|c} \mathcal{A}[T_U]_{\mathcal{A}} & 0 \\ \hline 0 & \mathcal{C}[T_W]_{\mathcal{C}} \end{array} \right]$$

$$= \mathrm{diag}(\mathfrak{b}(a_1, s_1), \mathfrak{b}(a_2, s_2), \ldots, \mathfrak{b}(a_r, s_r))$$

Thus the result follows in this case.

Now suppose that V is indecomposable with respect to T. Since F is algebraically closed, we can choose $a \in F$ to be an eigenvalue for T and set $S = T - aI$. By definition S is a singular linear transformation. Let $U = VS^n$ and $W = (0)\overleftarrow{S^n}$ be as in the preceding lemma. Then U and W are both S-invariant subspaces with $V = U \oplus W$. We now observe that U and W are T-invariant. To this end, suppose that $\alpha \in U$ and $\gamma \in W$. Since $\alpha S \in U$ and $\gamma S \in W$, we have

$$\alpha T = \alpha(S + aI) = \alpha S + a\alpha \in U$$
$$\gamma T = \gamma(S + aI) = \gamma S + a\gamma \in W$$

Therefore $V = U \oplus W$ is a direct sum of T-invariant subspaces.

But V is indecomposable with respect to T so we conclude that either $U = 0$ and $V = W$ or $U = V$ and $W = 0$. By the preceding lemma, S_U is nonsingular. Since we know that $S\colon V \to V$ is singular, we cannot have $U = V$. Thus $U = 0$ and $V = W$ and therefore $S = S_W$ is nilpotent. We have shown that $S = T - aI$ is nilpotent and thus, by Corollary 20.1, there exists a basis $\mathcal{B}$ of V such that

$$_\mathcal{B}[T]_\mathcal{B} = \operatorname{diag}(\mathfrak{b}(a, s_1), \mathfrak{b}(a, s_2), \ldots, \mathfrak{b}(a, s_r))$$

with $s_1 \geq s_2 \geq \cdots \geq s_r$. The theorem is proved. □

We remark that the above Jordan blocks for T are in fact unique up to the order in which they occur. However, we will not prove this. Also if $_\mathcal{B}[T]_\mathcal{B}$ is written as above, then the elements of $\mathcal{B}$ have a rather nice property. Indeed, if $\beta \in \mathcal{B}$ then there exists $a \in F$ with $\beta(T - aI)^m = 0$ for some $m \geq 1$. This property characterizes what are called *generalized eigenvectors* of T. Therefore an immediate consequence of the above theorem is

COROLLARY 21.1. *Let V be a finite dimensional vector space over an algebraically closed field F. If $T\colon V \to V$ is a linear transformation, then V has a basis $\mathcal{B}$ consisting of generalized eigenvectors of T.*

This concludes our work on finding a basis $\mathcal{B}$ that makes the matrix $_\mathcal{B}[T]_\mathcal{B}$ as simple as possible. We have obviously achieved our goal in case the field F is algebraically closed, but we have done little or nothing for nonalgebraically closed fields. Nevertheless a theory does exist in this more general situation and one gets the so-called *rational canonical form*.

The key to this form is a closer study of the polynomial ring $F[x]$. In the algebraically closed case, every polynomial $f(x) \in F[x]$ can be factored as

$$f(x) = b(x - a_1)(x - a_2) \cdots (x - a_n)$$

with b and the a_i in F. However this is no longer true in general as the examples

$$f(x) = x^2 + 1 \in \mathbb{R}[x]$$
$$g(x) = x^2 - 2 \in \mathbb{Q}[x]$$

show. Even so, one can prove that $F[x]$ has divisibility and factorization properties very similar to those of the ordinary integers.

If $f(x)$ is a monic polynomial in $F[x]$, let $c(f)$ denote its companion matrix. We recall that if $f(x) = x^s$ then

$$\mathfrak{b}(0, s) = c(x^s)$$

Now suppose that $T: V \to V$ is a linear transformation with characteristic polynomial $\varphi_T(x) = x^n$. Then our main result on nilpotent transformations, Theorem 20.1, states that there exists a basis $\mathcal{B}$ such that

$$_\mathcal{B}[T]_\mathcal{B} = \mathrm{diag}(c(x^{s_1}), c(x^{s_2}), \ldots, c(x^{s_r}))$$

where diag has the obvious meaning. Of course

$$x^{s_1} x^{s_2} \cdots x^{s_r} = x^n = \varphi_T(x)$$

This is precisely what happens in general in the rational canonical form. Let T be given. We can then show that there exists a basis $\mathcal{B}$ for V such that

$$_\mathcal{B}[T]_\mathcal{B} = \mathrm{diag}(c(f_1), c(f_2), \ldots, c(f_r))$$

where the f_i are monic polynomials with

$$\varphi_T(x) = f_1 f_2 \cdots f_r$$

We can further restrict the f_i to look even more like x^{s_i}. Namely the f_i can be taken to be powers of irreducible polynomials in $F[x]$, that is polynomials that cannot be factored in $F[x]$ as a product of polynomials of smaller degree.

Problems

Let V be a finite dimensional vector space over the algebraically closed field F and let $T: V \to V$. For each $a \in F$ let

$$V_a = \{\alpha \in V \mid \alpha(T - aI)^n = 0 \quad \text{for some} \quad n \geq 1\}$$

21.1. Prove that V_a is a subspace of V invariant under T. (V_a is the space of generalized eigenvectors for T with eigenvalue a.)

21.2. Show that $V_a \neq 0$ if and only if a is an eigenvalue of T.

21.3. Let $f(x)$ be a nonzero polynomial in $F[x]$ and let $a \in F$ with $f(a) \neq 0$. Suppose $\alpha \in V$ satisfies

$$\alpha f(T) = 0 \quad \text{and}$$

$$\alpha(T - aI)^n = 0 \quad \text{for some} \quad n \geq 1$$

Show that $\alpha = 0$. (Hint. First prove this for $n = 1$. In general, set $\beta = \alpha(T - aI)^{n-1}$ and consider $\beta f(T)$ and $\beta(T - aI)$.)

21.4. Let $\alpha \in V_{a_1} + V_{a_2} + \cdots + V_{a_r}$. Show that there exist integers $n_1, n_2, \ldots, n_r \geq 1$ with

$$\alpha(T - a_1I)^{n_1}(T - a_2I)^{n_2}\cdots(T - a_r)^{n_r} = 0$$

21.5. Use the above and Corollary 21.1 to conclude that

$$V = V_{a_1} \oplus V_{a_2} \oplus \cdots \oplus V_{a_t}$$

where $a_1, a_2, \ldots, a_t$ are the distinct eigenvalues of T.

21.6. If $k_i = \dim_F V_{a_i}$ show that

$$\varphi_T(x) = \prod_i (x - a_i)^{k_i}$$

Lemma 21.2 can be generalized to infinite dimensional vector spaces under suitable circumstances. Let V be an arbitrary vector space over F and let $T\colon V \to V$ be a linear transformation. Suppose there exists an integer $n \geq 1$ with

$$U = (V)T^n = (V)T^{2n}$$
$$W = (0)\overleftarrow{T^n} = (0)\overleftarrow{T^{2n}}$$

21.7. Show that U and W are both T-invariant subspaces of V and that T_U is onto and T_W is nilpotent.

21.8. Prove that $V = U + W$.

21.9. Prove that $U \cap W = 0$ so that $V = U \oplus W$. (Hint. Let $\alpha \in U \cap W$. Since $\alpha \in U = UT^n$ we have $\alpha = \beta T^n$. Then $0 = \alpha T^n = \beta T^{2n}$ and we can deduce information about β.)

21.10. Show that T_U is one-to-one and hence nonsingular.

CHAPTER IV

Bilinear Forms

22. Bilinear Forms

There are many other aspects of linear algebra. The one we will consider for the remainder of this course has its roots in geometry, in the study of conic sections. Let us consider the real Euclidean plane $\mathbb{R}^2$ with (x, y)-coordinate axes. Then the conic sections are graphs of equations of the form

$$ax^2 + 2bxy + cy^2 + dx + ey + f = 0$$

As one sees in analytic geometry, the main interest here is in the quadratic part

$$Q(x, y) = ax^2 + 2bxy + cy^2$$

and this is a function from $\mathbb{R}^2$ into the real numbers $\mathbb{R}$.

On the other hand, since this is a course in linear algebra, Q must somehow be related to a linear function. It is indeed. Suppose $\alpha_1 = (x_1, y_1)$ and $\alpha_2 = (x_2, y_2)$ are vectors in $\mathbb{R}^2$. We define

$$B(\alpha_1, \alpha_2) = ax_1x_2 + bx_1y_2 + by_1x_2 + cy_1y_2$$

and we now have a new function, this time from $\mathbb{R}^2 \times \mathbb{R}^2$ to $\mathbb{R}$. Now what happens when we set $\alpha_1 = \alpha_2 = \alpha = (x, y)$? Then

$$B(\alpha, \alpha) = ax^2 + 2bxy + cy^2 = Q(x, y)$$

and we are back where we started.

Why then do we introduce B? The reason is that it is bilinear. Let α_1, α_2 be as above and set $\alpha_2' = (x_2', y_2')$. Then

$$
\begin{aligned}
B(\alpha_1, \alpha_2 + \alpha_2') &= ax_1(x_2 + x_2') + bx_1(y_2 + y_2') \\
&\quad + by_1(x_2 + x_2') + cy_1(y_2 + y_2') \\
&= (ax_1x_2 + bx_1y_2 + by_1x_2 + cy_1y_2) \\
&\quad + (ax_1x_2' + bx_1y_2' + by_1x_2' + cy_1y_2') \\
&= B(\alpha_1, \alpha_2) + B(\alpha_1, \alpha_2')
\end{aligned}
$$

Moreover if $r \in \mathbb{R}$ then

$$
\begin{aligned}
B(\alpha_1, r\alpha_2) &= ax_1(rx_2) + bx_1(ry_2) + by_1(rx_2) + cy_1(ry_2) \\
&= r(ax_1x_2 + bx_1y_2 + by_1x_2 + cy_1y_2) \\
&= rB(\alpha_1, \alpha_2)
\end{aligned}
$$

Of course a similar result holds if we study B as a function of its first variable.

Let V be a vector space over a field F. A *bilinear form* is a map

$$B \colon V \times V \to F$$

such that B is a linear transformation when viewed as a function of each of the two variables individually. Let us consider some examples.

EXAMPLE 22.1. We always have the zero map, namely

$$B(\alpha, \beta) = 0 \qquad \text{for all} \quad \alpha, \beta \in V$$

EXAMPLE 22.2. If $\dim_F V = 2$, then

$$\det: V \times V \to F$$

is certainly bilinear.

EXAMPLE 22.3. Let $V = F[x]$ and for each $a \in F$ let E_a denote the evaluation map given by $\alpha E_a = \alpha(a)$. For any $a, b \in F$ we can define $B_{a,b}$ by

$$B_{a,b}(\alpha, \beta) = (\alpha E_a)(\beta E_b) = \alpha(a)\beta(b)$$

EXAMPLE 22.4. Let V be the real vector space of all continuous real valued functions on the interval $0 \le x \le 1$. Let $\omega(x)$ be some fixed member of V and define

$$B(\alpha, \beta) = \int_0^1 \omega(t)\alpha(t)\beta(t)\, dt$$

In Fourier analysis, ω is usually called the *weight function*.

EXAMPLE 22.5. Let V be as above and now let $\kappa(x, y)$ be some fixed continuous real valued function defined on the unit square $0 \le x \le 1$, $0 \le y \le 1$. Then we can set

$$B(\alpha, \beta) = \int_0^1 \int_0^1 \kappa(u, v)\alpha(u)\beta(v)\, du\, dv$$

Here κ is usually called the *kernel function*.

In our study of bilinear forms, we will be concerned with the equation $B(\alpha, \beta) = 0$. Suppose α, β is a solution of the above so that $B(\alpha, \beta) = 0$. Is it necessarily true that $B(\beta, \alpha) = 0$? Definitely not. For example, let V and $B_{a,b}$ be as in Example 22.3 with $a \ne b$. If $\alpha = 1$ and $\beta = x - b$, then

$$B_{a,b}(\alpha, \beta) = 1 \cdot (b - b) = 0$$

but

$$B_{a,b}(\beta, \alpha) = (a - b) \cdot 1 \ne 0$$

In this unpleasant situation, the theory just does not work well. Therefore, we will usually restrict our attention to *normal bilinear forms*, that
is bilinear forms B satisfying

$$B(\alpha, \beta) = 0 \quad \text{if and only if} \quad B(\beta, \alpha) = 0$$

There are two important special cases here.
    First, we say that a bilinear form B is *symmetric* if

$$B(\alpha, \beta) = B(\beta, \alpha) \qquad \text{for all} \quad \alpha, \beta \in V$$

Obviously such a form is also normal. Second, we say that B is *skew-
symmetric* if

$$B(\alpha, \alpha) = 0 \qquad \text{for all} \quad \alpha \in V$$

The following lemma explains the name and shows that these forms
are also normal.

LEMMA 22.1. *Let B be a skew-symmetric bilinear form on V. Then
for all $\alpha, \beta \in V$ we have*

$$B(\beta, \alpha) = -B(\alpha, \beta)$$

Hence B is normal.

PROOF. We have $B(\alpha + \beta, \alpha + \beta) = 0$ and therefore

$$\begin{aligned}
0 = B(\alpha + \beta, \alpha + \beta) &= B(\alpha, \alpha + \beta) + B(\beta, \alpha + \beta) \\
&= B(\alpha, \alpha) + B(\alpha, \beta) + B(\beta, \alpha) + B(\beta, \beta) \\
&= B(\alpha, \beta) + B(\beta, \alpha)
\end{aligned}$$

since also $B(\alpha, \alpha) = B(\beta, \beta) = 0$. Thus $B(\beta, \alpha) = -B(\alpha, \beta)$ and this
clearly implies that B is normal.   □

We now see that these are not only some examples of normal forms,
but they are in fact the only examples.

THEOREM 22.1. *Let $B \colon V \times V \to F$ be a normal bilinear form.
Then B is either symmetric or skew-symmetric.*

PROOF. Let $\sigma, \tau, \eta \in V$. Then by linearity

$$\begin{aligned}
B\big(\sigma, B(\sigma, \eta)\tau &- B(\sigma, \tau)\eta\big) \\
&= B(\sigma, \eta)(B(\sigma, \tau) - B(\sigma, \tau)B(\sigma, \eta) = 0
\end{aligned}$$

Hence, since B is normal, we have

$$\begin{aligned}
0 &= B\big(B(\sigma, \eta)\tau - B(\sigma, \tau)\eta, \sigma\big) \\
&= B(\sigma, \eta)B(\tau, \sigma) - B(\sigma, \tau)B(\eta, \sigma)
\end{aligned}$$

and we have shown that

$$B(\sigma, \eta)B(\tau, \sigma) = B(\eta, \sigma)B(\sigma, \tau) \qquad (*)$$

for all $\sigma, \tau, \eta \in V$.

Let us suppose that B is not skew-symmetric. Then we can choose a fixed $\gamma \in V$ with $B(\gamma, \gamma) \neq 0$. If $\alpha \in V$, then setting $\sigma = \eta = \gamma$, $\tau = \alpha$ in $(*)$ yields

$$B(\gamma, \gamma)B(\alpha, \gamma) = B(\gamma, \gamma)B(\gamma, \alpha)$$

so since $B(\gamma, \gamma) \neq 0$ we have $B(\alpha, \gamma) = B(\gamma, \alpha)$ for all $\alpha \in V$.

Now let $\alpha, \beta \in V$. Since $B(\gamma, \gamma) \neq 0$ and $|F| \geq 2$, we can certainly find $a \in F$ with

$$B(\alpha + a\gamma, \gamma) = B(\alpha, \gamma) + aB(\gamma, \gamma) \neq 0$$

Then setting $\sigma = \alpha + a\gamma$, $\tau = \beta$ and $\eta = \gamma$ in $(*)$ yields

$$B(\alpha + a\gamma, \gamma)B(\beta, \alpha + a\gamma) = B(\gamma, \alpha + a\gamma)B(\alpha + a\gamma, \beta)$$

But as we have seen before

$$B(\gamma, \alpha + a\gamma) = B(\alpha + a\gamma, \gamma) \neq 0$$

so this common factor cancels and we obtain

$$B(\beta, \alpha + a\gamma) = B(\alpha + a\gamma, \beta)$$

Finally

$$B(\beta, \alpha + a\gamma) = B(\beta, \alpha) + aB(\beta, \gamma) \quad \text{and}$$
$$B(\alpha + a\gamma, \beta) = B(\alpha, \beta) + aB(\gamma, \beta)$$

so since $B(\beta, \gamma) = B(\gamma, \beta)$ we conclude that $B(\beta, \alpha) = B(\alpha, \beta)$. Since this is true for all $\alpha, \beta \in V$, B is a symmetric bilinear form and the result follows. $\qquad\square$

Let $B \colon V \times V \to F$ be a normal bilinear form. If $\alpha, \beta \in V$, we say that α and β are *perpendicular* and write $\alpha \perp \beta$ if $B(\alpha, \beta) = 0$. Obviously $\alpha \perp \beta$ if and only if $\beta \perp \alpha$. If S is a nonempty subset of V, we write

$$S^{\perp} - \{\beta \in V \mid \beta \perp \alpha \quad \text{for all} \quad u \in S\}$$

We now list a number of trivial observations.

LEMMA 22.2. *Let* $B \colon V \times V \to F$ *be a normal bilinear form and let* S, S_1 *and* S_2 *be nonempty subsets of* V. *Then*

   i. $S^{\perp}$ is a subspace of V.
   ii. $S \subseteq S^{\perp\perp}$.
   iii. $S_1 \subseteq S_2$ implies that $S_2^{\perp} \subseteq S_1^{\perp}$.
   iv. $(S_1 \cup S_2)^{\perp} = S_1^{\perp} \cap S_2^{\perp}$.

    v. $\langle S \rangle^{\perp} = S^{\perp}$.

    PROOF. If $\alpha \in V$, then $\{\alpha\}^{\perp}$ is clearly the kernel of the linear transformation $\beta \mapsto B(\alpha, \beta)$ and is therefore a subspace of V. Since clearly

$$S^{\perp} = \bigcap_{\alpha \in S} \{\alpha\}^{\perp}$$

we see that $S^{\perp}$ is a subspace. This yields (i). Parts (ii), (iii) and (iv) are of course obvious.

    Now $\langle S \rangle \supseteq S$ so by (iii) we have $S^{\perp} \supseteq \langle S \rangle^{\perp}$. To obtain the reverse inclusion, we note that $S^{\perp\perp} \supseteq S$ and since $S^{\perp\perp}$ is a subspace, we must have $S^{\perp\perp} \supseteq \langle S \rangle$. This says that every element of $S^{\perp}$ is perpendicular to every element of $\langle S \rangle$ and hence $S^{\perp} \subseteq \langle S \rangle^{\perp}$. This yields (v).   □

    Finally we show

    THEOREM 22.2. *Let V be a finite dimensional vector space over F and let $B \colon V \times V \to F$ be a normal bilinear form. If W is a subspace of V, then*

$$\dim_F W + \dim_F W^{\perp} \geq \dim_F V$$

In particular, if $W \cap W^{\perp} = 0$, then $V = W \oplus W^{\perp}$.

    PROOF. If $W = 0$, then $W^{\perp} = V$ and the result is clear. Now suppose $\dim_F W = s \geq 1$ and let $S = \{\alpha_1, \alpha_2, \ldots, \alpha_s\}$ be a basis for W. Then by the previous lemma, $W^{\perp} = S^{\perp}$. Now $W^{\perp} = S^{\perp}$ is certainly the kernel of the linear transformation $T \colon V \to F^s$ given by $\beta \mapsto \Big(B(\alpha_1, \beta), B(\alpha_2, \beta), \ldots, B(\alpha_s, \beta) \Big)$. Hence since

$$\dim_F \operatorname{im} T \leq \dim_F F^s = s = \dim_F W$$

Theorem 9.3 yields

$$\dim_F V = \dim_F \operatorname{im} T + \dim_F \ker T$$

$$\leq \dim_F W + \dim_F W^{\perp}$$

as required.

    Finally if $W \cap W^{\perp} = 0$, then $W + W^{\perp} = W \oplus W^{\perp}$ is a subspace of V of dimension $\dim_F W + \dim_F W^{\perp} \geq \dim_F V$, so $W \oplus W^{\perp} = V$ and the theorem is proved.   □

Problems

    22.1. Suppose we defined a skew-symmetric bilinear form by the relation $B(\alpha, \beta) = -B(\beta, \alpha)$. Would we get the same answer?

22.2. Consider the bilinear form defined in Example 22.5. What do you think are the appropriate assumptions to make on the kernel function $\kappa(x, y)$ to guarantee that B is symmetric or skew-symmetric?

22.3. Discuss how one might define a multilinear form in n variables on V. What would the corresponding definitions be for symmetric or skew-symmetric forms?

22.4. Suppose $n = \dim_F V$. What are the skew-symmetric multilinear forms in n variables on V? Can you think of a symmetric analog?

22.5. Let $V = F[x]$ and define $B \colon V \times V \to F$ by
$$B(\alpha, \beta) = \alpha(0)\beta(0)$$
For each $\alpha \in V$ find $\{\alpha\}^{\perp}$. What is $V^{\perp}$?

22.6. Give an example to show that $\dim_F W + \dim_F W^{\perp}$ could be strictly larger than $\dim_F V$.

If $B, B' \colon V \times V \to F$ are bilinear forms, define $B + B'$ by
$$(B + B')(\alpha, \beta) = B(\alpha, \beta) + B'(\alpha, \beta)$$
Similarly, for $a \in F$ define aB by
$$(aB)(\alpha, \beta) = a \cdot B(\alpha, \beta)$$

22.7. Show that $B + B'$ and aB are bilinear forms.

22.8. If B and B' are both symmetric (or skew-symmetric) prove that the same is true for $B + B'$ and aB. What can one say about the sum of two normal forms?

22.9. Define $B^t \colon V \times V \to F$ by $B^t(\alpha, \beta) = B(\beta, \alpha)$. Discuss the bilinear forms $B + B^t$ and $B - B^t$.

22.10. Assume that F is a field with $2 = 1 + 1 \neq 0$. Prove that every bilinear form is a sum of a symmetric and a skew-symmetric form and uniquely so. What happens when $1 + 1 = 0$?

23. Symmetric and Skew-Symmetric Forms

Let $B\colon V \times V \to F$ be a normal bilinear form on the vector space V over F. If W is a subspace of V, then it is clear that B defines, by restriction, a normal bilinear form $B_W\colon W \times W \to F$. That is, for all $\alpha, \beta \in W$, $B_W(\alpha, \beta)$ is just $B(\alpha, \beta)$. It is necessary now to introduce a certain amount of notation.

First, it is quite possible that B_W is the zero form. That is, for all $\alpha, \beta \in W$

$$B_W(\alpha, \beta) = B(\alpha, \beta) = 0$$

In this case, we say that W is an *isotropic* subspace of V. Otherwise W is said to be *nonisotropic*.

Now there is a natural way of finding a particular isotropic subspace. We define the *radical* of V to be

$$\operatorname{rad} V = \operatorname{rad}_B V = V^\perp$$

Observe that if $\alpha \in V^\perp$ and $\beta \in V$, then $B(\alpha, \beta) = 0$ and therefore all the more so we have $B(\alpha, \beta) = 0$ if $\beta \in V^\perp$. Thus we see that the radical of V is an example of an isotropic subspace.

If $\operatorname{rad} V = V^\perp = 0$ we say that V is *nonsingular*. In the same way, if W is a subspace of V then W is nonsingular if W is nonsingular with respect to the restricted bilinear form $B_W\colon W \times W \to F$. Clearly

$$\operatorname{rad}_{B_W} W = W \cap W^\perp$$

where $W^\perp$ is defined in V. Thus we see that W is a nonsingular subspace if and only if $W \cap W^\perp = 0$. Let us consider some examples.

EXAMPLE 23.1. Suppose B is a symmetric form and assume that for some $\alpha \in V$ we have $B(\alpha, \alpha) \neq 0$. Then certainly $\alpha \neq 0$ so $W = \langle \alpha \rangle$ is a one-dimensional subspace of V. If $\beta, \gamma \in W$ are both nonzero, then $\beta = b\alpha$ and $\gamma = c\alpha$ for some nonzero $b, c \in F$ and thus

$$B_W(\beta, \gamma) = B(b\alpha, c\alpha) = bcB(\alpha, \alpha) \neq 0$$

We conclude that W is a nonisotropic and in fact a nonsingular subspace of V. Using the geometric designation for one-dimensional spaces, we call such a subspace W a *nonisotropic line*.

EXAMPLE 23.2. Now let B be skew-symmetric and let $U = \langle \alpha \rangle$ be a one-dimensional subspace of V. If $\beta, \gamma \in U$ then $\beta = b\alpha$, $\gamma = c\alpha$ for some $b, c \in F$ and thus

$$B_U(\beta, \gamma) = B(b\alpha, c\alpha) = bcB(\alpha, \alpha) = 0$$

since B is skew-symmetric. This shows that U is an isotropic subspace and therefore that a skew-symmetric form does not give rise to nonisotropic lines. However, we do get interesting nonisotropic planes, that is nonisotropic 2-dimensional subspaces.

Let W be a 2-dimensional subspace of V with basis $\{\alpha, \beta\}$ and suppose that

$$B(\alpha, \alpha) = 0, \qquad B(\alpha, \beta) = 1$$
$$B(\beta, \alpha) = -1, \qquad B(\beta, \beta) = 0$$

Then we say that W is a *hyperbolic plane*. We observe now that W is nonsingular. Let $\gamma \in W$ and write $\gamma = a\alpha + b\beta$ with $a, b \in F$. Then by the above

$$B(\alpha, \gamma) = B(\alpha, a\alpha + b\beta)$$
$$= aB(\alpha, \alpha) + bB(\alpha, \beta) = b$$

and

$$B(\beta, \gamma) = B(\beta, a\alpha + b\beta)$$
$$= aB(\beta, \alpha) + bB(\beta, \beta) = -a$$

Thus if $\gamma \in \operatorname{rad} W = W \cap W^\perp$, then $a = b = 0$ and $\gamma = 0$.

We see below that the above two examples are typical of the non-singular subspaces that exist. In dealing with symmetric forms it is necessary to eliminate certain fields from consideration. Namely we must assume that $2 = 1 + 1 \neq 0$ in F.

LEMMA 23.1. *Let B be a normal bilinear form on the vector space V and assume that V is nonisotropic.*

> i. *If B is symmetric and $1 + 1 \neq 0$ in F, then V contains a nonisotropic line.*
> ii. *If B is skew-symmetric, then V contains a hyperbolic plane.*

Furthermore, these subspaces are nonsingular.

PROOF. (*i*) Here we assume that B is symmetric. Suppose first that there exists $\alpha \in V$ with $B(\alpha, \alpha) \neq 0$. Then as we have seen in Example 23.1, $W = \langle \alpha \rangle$ is a nonisotropic line and a nonsingular subspace of V.

We must now consider the case in which $B(\alpha, \alpha) = 0$ for all $\alpha \in V$. Then by definition, B is also skew-symmetric and by Lemma 22.1 we have $B(\alpha, \beta) = -B(\beta, \alpha)$ for all $\alpha, \beta \in V$. On the other hand, B is symmetric so

$$B(\alpha, \beta) = -B(\beta, \alpha) = -B(\alpha, \beta)$$

and $2B(\alpha, \beta) = 0$. Since $2 \neq 0$ in F we conclude that $B(\alpha, \beta) = 0$ for all $\alpha, \beta \in V$ and hence V is isotropic, a contradiction.

(ii) Here B is skew-symmetric and V is nonisotropic so we can choose $\alpha, \beta \in V$ with $B(\alpha, \beta) \neq 0$. Certainly α and β are nonzero and if $b \in F$ then $B(\alpha, b\beta) = bB(\alpha, \beta)$. We can conclude two facts from this latter observation. First by replacing β by some $b\beta$ if necessary we can assume that $B(\alpha, \beta) = 1$. Second, we see that α and β are linearly independent since otherwise $\alpha = b\beta$ for some $0 \neq b \in F$ and then

$$B(\alpha, \alpha) = B(\alpha, b\beta) = bB(\alpha, \beta) \neq 0$$

which contradicts the fact that B is skew-symmetric. Therefore $W = \langle \alpha, \beta \rangle$ is 2-dimensional and we may assume that $B(\alpha, \beta) = 1$. Since B is skew-symmetric we then have $B(\alpha, \alpha) = 0$, $B(\beta, \beta) = 0$ and $B(\beta, \alpha) = -B(\alpha, \beta) = -1$. Thus W is a hyperbolic plane. Finally as we saw in Example 23.2, such planes are always nonsingular and the lemma is proved.   □

Of course these lines and planes are rather small subspaces of V. We now consider a way to build larger subspaces. Let U, W be subspaces of V. We say that U and W are perpendicular and write $U \perp W$ if for all $\alpha \in U$, $\beta \in W$ we have $\alpha \perp \beta$. Clearly $U \perp W$ if and only if $U \subseteq W^\perp$ or equivalently $W \subseteq U^\perp$. Now let $W_1, W_2, \ldots, W_k$ be subspaces of V. We say that $W_1 + W_2 + \cdots + W_k$ is a *perpendicular direct sum* if first of all it is a direct sum and secondly if for all $i \neq j$ we have $W_i \perp W_j$.

LEMMA 23.2. *Suppose $B \colon V \times V \to F$ is a normal bilinear form and let*

$$W = W_1 + W_2 + \cdots + W_k$$

be a perpendicular direct sum. If each W_i is nonsingular, then so is W.

PROOF. Let $\alpha = \alpha_1 + \alpha_2 + \cdots + \alpha_k \in \operatorname{rad} W$ with each $\alpha_i \in W_i$. We will show that each α_j is zero. To this end, fix j and let β_j be any element of W_j. Since $W_j \subseteq W$, we have $B(\alpha, \beta_j) = 0$. Now

$$\begin{aligned} 0 = B(\alpha, \beta_j) &= B(\alpha_1 + \alpha_2 + \cdots + \alpha_k, \beta_j) \\ &= B(\alpha_1, \beta_j) + B(\alpha_2, \beta_j) + \cdots + B(\alpha_k, \beta_j) \end{aligned}$$

and thus since $W_i \perp W_j$ for all $i \neq j$ we have $B(\alpha_j, \beta_j) = 0$ for all $\beta_j \in W_j$. Therefore $\alpha_j \in \operatorname{rad} W_j = 0$ and, since this is true for all α_j, we have $\alpha = 0$ and W is nonsingular.   □

We now come to the main result of this section.

THEOREM 23.1. *Let V be a finite dimensional vector space over F and let $B : V \times V \to F$ be a normal bilinear form.*

    i. If B is symmetric and if $1 + 1 \neq 0$ in F, then

$$V = W_1 + W_2 + \cdots + W_k + U$$

    can be written as a perpendicular direct sum where the subspaces W_i are nonisotropic lines and U is isotropic.

    ii. If B is skew-symmetric, then

$$V = W_1 + W_2 + \cdots + W_k + U$$

    is a perpendicular direct sum where the subspaces W_i are now hyperbolic planes and U is an isotropic subspace.

PROOF. We prove both parts simultaneously. Let us consider all subspaces W of V that can be written as

$$W = W_1 + W_2 + \cdots + W_k$$

a perpendicular direct sum of the subspaces W_i. Here W_i is a nonisotropic line in case B is symmetric or W_i is a hyperbolic plane in case B is skew-symmetric. Certainly $W = 0$ is such a subspace if we view the empty sum as 0.

Now among all such subspaces let us choose one, say W as given above, of maximal dimension. By Lemmas 23.1 and 23.2 we see that W is a nonsingular subspace and thus by Theorem 22.2 we have $V = W \oplus W^\perp$. Suppose $W^\perp$ is a nonisotropic subspace of V. Then by Lemma 23.1 again, there exists a subspace $W_{k+1} \subseteq W^\perp$ that is either a nonisotropic line if B is symmetric or a hyperbolic plane if B is skew-symmetric. But then $W \cap W_{k+1} = 0$ and $W_{k+1} \subseteq W^\perp$ imply that

$$W' = W \oplus W_{k+1} = W_1 + W_2 + \cdots + W_k + W_{k+1}$$

is an appropriate perpendicular direct sum with $W' > W$ and hence $\dim_F W' > \dim_F W$, a contradiction. Therefore we conclude that $W^\perp$ is isotropic and since $W \oplus W^\perp$ is a perpendicular direct sum we have

$$V = W \oplus W^\perp = W_1 + W_2 + \cdots + W_k + W^\perp$$

and the result follows. $\qquad\square$

It is interesting to observe that the above result essentially determines B. First let B be symmetric and let V be decomposed as above. Let $W_i = \langle \delta_i \rangle$ and set $d_i = B(\delta_i, \delta_i)$. If $\alpha, \beta \in V$, then we can write these elements uniquely as

$$\alpha = a_1 \delta_1 + a_2 \delta_2 + \cdots + a_k \delta_k + \alpha'$$
$$\beta = b_1 \delta_1 + b_2 \delta_2 + \cdots + b_k \delta_k + \beta'$$

with $\alpha', \beta' \in U$. From the perpendicularity of the sum for V, we have easily

$$B(\alpha, \beta) = a_1 b_1 B(\delta_1, \delta_1) + a_2 b_2 B(\delta_2, \delta_2) + \cdots + a_k b_k B(\delta_k, \delta_k)$$
$$+ B(\alpha', \beta')$$
$$= a_1 b_1 d_1 + a_2 b_2 d_2 + \cdots + a_k b_k d_k$$

Similarly if B is skew-symmetric, let $W_i = \langle \gamma_i, \delta_i \rangle$ with $B(\gamma_i, \delta_i) = 1$, $B(\delta_i, \gamma_i) = -1$. If $\alpha, \beta \in V$, then we can write

$$\alpha = \sum_{i=1}^{k} (a_i \gamma_i + \tilde{a}_i \delta_i) + \alpha'$$

$$\beta = \sum_{i=1}^{k} (b_i \gamma_i + \tilde{b}_i \delta_i) + \beta'$$

with $\alpha', \beta' \in U$. Again the perpendicularity of the sum for V yields

$$B(\alpha, \beta) = \sum_{i=1}^{k} B(a_i \gamma_i + \tilde{a}_i \delta_i, b_i \gamma_i + \tilde{b}_i \delta_i) + B(\alpha', \beta')$$

$$= \sum_{i=1}^{k} (a_i \tilde{b}_i - \tilde{a}_i b_i)$$

Problems

Let V be a vector space over F and let $B \colon V \times V \to F$ be a normal bilinear form.

23.1. Let S be a nonempty subset of V. Show that $S^{\perp\perp\perp} = S^{\perp}$.

23.2. Let $V = W_1 + W_2 + \cdots + W_k$ be a perpendicular direct sum. Prove that

$$\operatorname{rad} V = \operatorname{rad} W_1 + \operatorname{rad} W_2 + \cdots + \operatorname{rad} W_k$$

23.3. In Theorem 23.1, show that the subspace U is in fact equal to $\operatorname{rad} V$.

23.4. Let $V = W_1 + W_2 + \cdots + W_k$ be a sum of subspaces with $W_i \perp W_j$ for all $i \neq j$ and with each W_i nonsingular. Prove that the above sum is direct.

23.5. Let F be a field with $1 + 1 = 0$. What is the difference between symmetric and skew-symmetric forms?

23.6. Again let F be a field with $1 + 1 = 0$ and suppose that B is symmetric. Find an appropriate decomposition for the finite dimensional vector space V.

A *quadratic form* on V is a map $Q \colon V \to F$ satisfying

   i. $Q(a\alpha) = a^2 Q(\alpha)$ for all $a \in F$ and $\alpha \in V$.

  ii. If $R(\alpha, \beta)$ is defined by

$$R(\alpha, \beta) = Q(\alpha + \beta) - Q(\alpha) - Q(\beta)$$

    then R is a bilinear form.

23.7. Let B be a bilinear form and set $Q(\alpha) = B(\alpha, \alpha)$. Prove that Q is a quadratic form. What is the corresponding bilinear form R?

23.8. Show that R is symmetric and $R(\alpha, \alpha) = 2Q(\alpha)$ for all $\alpha \in V$.

23.9. Suppose F is a field with $1 + 1 \neq 0$. Show that $Q \leftrightarrow R$ is a one-to-one correspondence between quadratic forms and symmetric bilinear forms.

23.10. Let V be finite dimensional and let $1 + 1 \neq 0$ in F. Using Theorem 23.1 show that there exists a basis $\langle \alpha_1, \alpha_2, \ldots, \alpha_n \}$ of V and field elements $d_1, d_2, \ldots, d_n$ so that if $\alpha = \sum a_i \alpha_i$, then $Q(\alpha) = \sum a_i^2 d_i$.

24. Congruent Matrices

So far bilinear forms have really just been abstract objects for us. As with linear transformations we can achieve a concrete realization by fixing a basis for the vector space. It turns out that we can also associate these forms in a rather natural way with a matrix.

Let $B\colon V \times V \to F$ be a bilinear form and let $\mathcal{C} = \{\gamma_1, \gamma_2, \ldots, \gamma_n\}$ be an ordered basis for V. Then for all i, j we have field elements $B(\gamma_i, \gamma_j)$ and, since this is a doubly subscripted system, it seems reasonable to make these the entries of a suitable $n \times n$ matrix. Thus we define the matrix of B with respect to $\mathcal{C}$ to be

$$[B]^{\mathcal{C}} = [B(\gamma_i, \gamma_j)]$$

THEOREM 24.1. *Let $\mathcal{C} = \{\gamma_1, \gamma_2, \ldots, \gamma_n\}$ be a fixed basis for V. Then the map $B \mapsto [B]^{\mathcal{C}}$ yields a one-to-one correspondence between the set of all bilinear forms B on V and the set $F^{n \times n}$ of all $n \times n$ matrices over F.*

PROOF. We first show that the map is onto. Let $[c_{ij}] \in F^{n \times n}$ and define a maps $B\colon V \times V \to F$ as follows. If $\alpha = \sum_i a_i \gamma_i$, $\beta = \sum_i b_i \gamma_i$ are vectors in V, then

$$B(\alpha, \beta) = \sum_{i,j} a_i c_{ij} b_j$$

Since the a's and b's occur linearly in this formula, it is clear that B is a bilinear form. Finally since $\gamma_i = 0\gamma_1 + 0\gamma_2 + \cdots + 1\gamma_i + \cdots + 0\gamma_n$ we have $B(\gamma_i, \gamma_j) = c_{ij}$ so this particular B maps to the matrix $[c_{ij}]$ and the map is onto.

We show now that the map is one-to-one or in other words that B is uniquely determined by $[B]^{\mathcal{C}}$. Let $\alpha, \beta \in V$ and write $\alpha = \sum_i a_i \gamma_i$ and $\beta = \sum_i b_i \gamma_i$. Then by the linear properties of B, we have

$$B(\alpha, \beta) = B\left(\sum_i a_i \gamma_i, \sum_j b_j \gamma_j\right)$$

$$= \sum_i a_i B\left(\gamma_i, \sum_j b_j \gamma_j\right)$$

$$= \sum_{i,j} a_i B(\gamma_i, \gamma_j) b_j$$

Now the a's and b's depend only upon α, β and $\mathcal{C}$ and the field elements $B(\gamma_i, \gamma_j)$ depend only upon $[B]^{\mathcal{C}}$. Thus we see that $\mathcal{C}$ and $[B]^{\mathcal{C}}$ uniquely determine B and the theorem is proved. $\qquad \square$

Let us record the above formula for later use. Thus if $B \colon V \times V \to F$ is a bilinear form and if $\alpha = \sum_i a_i \gamma_i$ and $\beta = \sum_i b_i \gamma_i$ then

$$B(\alpha, \beta) = \sum_{i,j} a_i B(\gamma_i, \gamma_j) b_j \qquad (*)$$

Now what sort of matrices do the symmetric and skew-symmetric forms correspond to? Recall that if $[c_{ij}]$ is a matrix then its transpose is defined to be $[c_{ij}]^\mathsf{T} = [d_{ij}]$ where $d_{ij} = c_{ji}$. In other words, the transpose of a matrix is essentially its mirror image about the main diagonal. Let $[c_{ij}] \in F^{n \times n}$. We say that $[c_{ij}]$ is a *symmetric matrix* if $[c_{ij}]^\mathsf{T} = [c_{ij}]$ or equivalently if $c_{ij} = c_{ji}$ for all i, j. We say that $[c_{ij}]$ is a *skew-symmetric matrix* if $[c_{ij}]^\mathsf{T} = -[c_{ij}]$ and if all the diagonal entries are zero or equivalently if $c_{ij} = -c_{ji}$ for all i, j and $c_{ii} = 0$ for all i. Of course we now have

THEOREM 24.2. *Under the correspondence $B \leftrightarrow [B]^{\mathcal{C}}$ symmetric bilinear forms correspond to symmetric matrices and skew-symmetric forms correspond to skew-symmetric matrices.*

PROOF. Let $\mathcal{C} = \{\gamma_1, \gamma_2, \ldots, \gamma_n\}$. Suppose first that B is symmetric. Then $B(\gamma_i, \gamma_j) = B(\gamma_j, \gamma_i)$ so $[B]^{\mathcal{C}} = B(\gamma_i, \gamma_j)]$ is a symmetric matrix. Conversely let us assume that the matrix is symmetric and let α and β be given as in equation $(*)$. Then by that equation we have

$$B(\alpha, \beta) = \sum_{i,j} a_i B(\gamma_i, \gamma_j) b_j$$

$$= \sum_{i,j} b_j B(\gamma_j, \gamma_i) a_i = B(\beta, \alpha)$$

and B is symmetric.

Now let B be skew-symmetric. Then by Lemma 22.1 we have $B(\gamma_i, \gamma_i) = 0$ and $B(\gamma_i, \gamma_j) = -B(\gamma_j, \gamma_i)$ and hence $[B(\gamma_i, \gamma_j)]$ is a skew-symmetric matrix. Finally let us suppose that the matrix is skew symmetric and let $\alpha = \beta$ in equation $(*)$. Then

$$B(\alpha, \alpha) = \sum_{i,j} a_i B(\gamma_i, \gamma_j) a_j$$

Since the diagonal entries of the matrix are 0, the terms with $i = j$ do not appear in the above sum. Thus we may assume that $i \neq j$ and then the contribution from the pair $\{i, j\}$ is

$$a_i B(\gamma_i, \gamma_j) a_j + a_j B(\gamma_j, \gamma_i) a_i$$

$$= a_i a_j \Big(B(\gamma_i, \gamma_j) + B(\gamma_j, \gamma_i) \Big) = 0$$

since the matrix is skew-symmetric. This shows that $B(\alpha, \alpha) = 0$ for all $\alpha \in V$ and hence B is skew-symmetric. □

Now as we saw at the end of the last section certain bases exist for V that make symmetric and skew-symmetric forms looks nice. It is certainly natural to ask how the matrix $[B]^{\mathcal{C}}$ changes under a change of basis. The answer is given below.

THEOREM 24.3. *Let* $B \colon V \times V \to F$ *be a bilinear form and let* $\mathcal{A}$ *and* $\mathcal{C}$ *be two ordered bases for the finite dimensional vector space* V. *Then*

$$[B]^{\mathcal{A}} = {}_{\mathcal{A}}[I]_{\mathcal{C}} \cdot [B]^{\mathcal{C}} \cdot {}_{\mathcal{A}}[I]_{\mathcal{C}}{}^{\mathsf{T}}$$

where ${}_{\mathcal{A}}[I]_{\mathcal{C}}$ *is the usual change of basis matrix.*

PROOF. Let $\mathcal{A} = \{\alpha_1, \alpha_2, \ldots, \alpha_n\}$ and $\mathcal{C} = \{\gamma_1, \gamma_2, \ldots, \gamma_n\}$. Recall that ${}_{\mathcal{A}}[I]_{\mathcal{C}} = [a_{ij}]$ means that $\alpha_i = \sum_j a_{ij}\gamma_j$. Then by equation $(*)$ we have

$$B(\alpha_r, \alpha_s) = \sum_{i,j} a_{ri} B(\gamma_i, \gamma_j) a_{sj}$$

and the right hand side is clearly the r, s-entry of the matrix product $[a_{ij}][B(\gamma_i, \gamma_j)][a_{ij}]^{\mathsf{T}} = {}_{\mathcal{A}}[I]_{\mathcal{C}} \cdot [B]^{\mathcal{C}} \cdot {}_{\mathcal{A}}[I]_{\mathcal{C}}{}^{\mathsf{T}}$. The result follows. □

Suppose that α and γ are matrices in $F^{n \times n}$. We say that α and γ are *congruent* if there exists a nonsingular matrix β with $\alpha = \beta\gamma\beta^{\mathsf{T}}$. As an immediate consequence of the above we have

LEMMA 24.1. *Let* $\alpha, \gamma \in F^{n \times n}$ *and let* V *be some* n-*dimensional vector space over* F, *for example* $V = F^n$. *Then* α *and* γ *are congruent matrices if and only if there exists a bilinear form* $B \colon V \times V \to F$ *and bases* $\mathcal{A}$ *and* $\mathcal{C}$ *with* $\alpha = [B]^{\mathcal{A}}$ *and* $\gamma = [B]^{\mathcal{C}}$.

PROOF. If $\alpha = [B]^{\mathcal{A}}$ and $\gamma = [B]^{\mathcal{C}}$ then by the preceding theorem, α and γ are certainly congruent. Conversely assume that $\alpha = \beta\gamma\beta^{\mathsf{T}}$ for some nonsingular matrix β. Let $\mathcal{C}$ be a fixed ordered basis for V. Then by Theorem 24.1 we see that there exists a bilinear form B with $[B]^{\mathcal{C}} = \gamma$. Now β is a nonsingular matrix so by Lemma 12.3 there exists a basis $\mathcal{A}$ of V with $\beta = {}_{\mathcal{A}}[I]_{\mathcal{C}}$. Thus

$$\alpha = \beta\gamma\beta^{\mathsf{T}} = {}_{\mathcal{A}}[I]_{\mathcal{C}} \cdot [B]^{\mathcal{C}} \cdot {}_{\mathcal{A}}[I]_{\mathcal{C}}{}^{\mathsf{T}} = [B]^{\mathcal{A}}$$

by the above theorem and the result follows. □

We can now combine all of the above ideas to prove some results about symmetric and skew-symmetric matrices.

THEOREM 24.4. *Let* α *be a matrix in* $F^{n \times n}$.

   i. *If α is symmetric and $1 + 1 \neq 0$ in F, then α is congruent to a diagonal matrix.*

   ii. *If α is skew-symmetric, then α is congruent to a block diagonal matrix of the form*

$$\mathrm{diag}(H, H, \ldots, H, Z)$$

where

$$H = \begin{bmatrix} 0 & 1 \\ -1 & 0 \end{bmatrix}$$

corresponds to the hyperbolic plane, and Z is a suitable zero matrix.

Specifically, the block diagonal matrix in (ii) above looks like

$$
\begin{bmatrix}
0 & 1 & & & & & & & \\
-1 & 0 & & & & & & & \\
& & 0 & 1 & & & & & \\
& & -1 & 0 & & & & & \\
& & & & \ddots & & & & \\
& & & & & 0 & 1 & & \\
& & & & & -1 & 0 & & \\
& & & & & & & 0 & \\
& & & & & & & & \ddots \\
& & & & & & & & & 0
\end{bmatrix}
$$

PROOF. Let V be some n-dimensional vector space over F, say for example $V = F^n$. Fix a basis $\mathcal{A}$ of V so that by Theorem 24.1, $\alpha = [B]^{\mathcal{A}}$ for some bilinear form B.

   (i) Suppose first that α is a symmetric matrix and $1 + 1 \neq 0$ in F. Then by Theorem 24.2, B is a symmetric bilinear form. Therefore by Theorem 23.1, we can write

$$V = W_1 + W_2 + \cdots + W_k + U$$

a perpendicular direct sum of the nonisotropic lines W_i and also the isotropic space U. Let $W_i = \langle \gamma_i \rangle$ and let $\{\gamma_{k+1}, \ldots, \gamma_n\}$ be a basis for U. Then $\mathcal{C} = \{\gamma_1, \ldots, \gamma_k, \ldots, \gamma_n\}$ is a basis for V and we consider $[B]^{\mathcal{C}}$. Let $i \neq j$. If either $i \leq k$ or $j \leq k$, then $B(\gamma_i, \gamma_j) = 0$ since the above is a perpendicular direct sum. On the other hand, if $i > k$ and $j > k$, then $\gamma_i, \gamma_j \in U$ so again $B(\gamma_i, \gamma_j) = 0$ since U is isotropic. Thus we shown that $[B]^{\mathcal{C}}$ has all zeros off the main diagonal. By Theorem 24.3, $\alpha = [B]^{\mathcal{A}}$ is congruent to the diagonal matrix $[B]^{\mathcal{C}}$.

   (ii) Now let α be a skew-symmetric matrix. Then by Theorem 24.2, B is a skew-symmetric bilinear form. Therefore by Theorem 23.1, we can write

$$V = W_1 + W_2 + \cdots + W_k + U$$

a perpendicular direct sum on the hyperbolic planes W_i and also the isotropic space U. Let $W_i = \langle \gamma_i, \delta_i \rangle$ with $B(\gamma_i, \delta_i) = 1$, $B(\delta_i, \gamma_i) = -1$ and let $\{\beta_1, \beta_2, \ldots, \beta_s\}$ be a basis for U. Then certainly

$$\mathcal{C} = \{\gamma_1, \delta_1, \gamma_2, \delta_2, \ldots, \gamma_k, \delta_k, \beta_1, \beta_2, \ldots, \beta_s\}$$

is a basis for V. We compute the matrix $[B]^{\mathcal{C}}$. Since the above is a perpendicular direct sum and since U is an isotropic subspace we have easily

$$B(\beta_i, \gamma_j) = 0 = B(\gamma_j, \beta_i)$$
$$B(\beta_i, \delta_j) = 0 = B(\delta_j, \beta_i)$$
$$B(\beta_i, \beta_j) = 0 = B(\beta_j, \beta_i)$$

for all i, j. Furthermore, for all $i \neq j$ we have $B(\gamma_i, \gamma_j) = 0 = B(\delta_i, \delta_j)$ and $B(\gamma_i, \delta_j) = 0 = B(\delta_j, \gamma_i)$. Thus the only nonzero entries of $[B]^{\mathcal{C}}$ occur when we consider two elements of each W_i. Here we have

$$B(\gamma_i, \gamma_i) = \quad 0 \quad B(\gamma_i, \delta_i) = 1$$
$$B(\delta_i, \gamma_i) = -1 \quad B(\delta_i, \delta_i) = 0$$

Therefore $[B]^{\mathcal{C}}$ has the indicated block diagonal form with precisely k blocks equal to H. Finally, by Theorem 24.3, $\alpha = [B]^{\mathcal{A}}$ is congruent to the matrix $[B]^{\mathcal{C}}$, as required. $\square$

Finally we should consider whether the matrices we get in the above theorem are actually unique. For symmetric matrices α we get diagonal matrices but the diagonal entries are by no means unique. For certain fields we can suitably restrict the entries to get a uniqueness theorem but the problem is certainly not solved in general. On the other hand, if α is skew-symmetric then the only parameter here is the number of H blocks that appear and this is in fact uniquely determined.

Problems

24.1. Let V be the real subspace of $\mathbb{R}[x]$ consisting of all polynomials of degree $< n$ and let $B \colon V \times V \to \mathbb{R}$ be defined by $B(\alpha, \beta) = \int_0^1 \alpha(x)\beta(x)\, dx$. If $\mathcal{C}$ is the basis of V given by $\{1, x, x^2, \ldots, x^{n-1}\}$ show that the matrix $[B]^{\mathcal{C}}$ is the Hilbert matrix of Example 16.2.

24.2. Prove that the map $B \colon V \times V \to F$ defined in the first paragraph of the proof of Theorem 24.1 is a bilinear form.

24.3. Let V be a vector space over F with basis $\mathcal{C} = \{\gamma_1, \gamma_2, \ldots, \gamma_n\}$ and let $B\colon V \times V \to F$ be a bilinear form. Let $\alpha = \sum_i a_i \gamma_i$ and $\beta = \sum_i b_i \gamma_i$ be vectors in V. Then the matrix product

$$[a_1\, a_2\, \ldots\, a_n]\cdot[B]^{\mathcal{C}}\cdot[b_1\, b_2\, \ldots\, b_n]^{\mathsf{T}}$$

is a 1×1 matrix. Show that its entry is precisely $B(\alpha, \beta)$.

24.4. Let V be a 3-dimensional $\mathbb{Q}$-vector space with basis $\mathcal{C} = \{\gamma_1, \gamma_2, \gamma_3\}$ and let $B\colon V \times V \to \mathbb{Q}$ be the symmetric bilinear form given by

$$[B]^{\mathcal{C}} = \begin{bmatrix} 0 & 1 & -2 \\ 1 & 2 & 1 \\ -2 & 1 & 3 \end{bmatrix}$$

Find a basis $\mathcal{A}$ such that $[B]^{\mathcal{A}}$ is diagonal.

24.5. Let V be a 4-dimensional $\mathbb{Q}$ vector space with basis $\mathcal{C} = \{\gamma_1, \gamma_2, \gamma_3, \gamma_4\}$ and let $B\colon V \times V \to \mathbb{Q}$ be the skew-symmetric bilinear form given by

$$[B]^{\mathcal{C}} = \begin{bmatrix} 0 & 1 & 2 & 3 \\ -1 & 0 & -1 & -2 \\ -2 & 1 & 0 & 0 \\ -3 & 2 & 0 & 0 \end{bmatrix}$$

Find a basis $\mathcal{A}$ such that $[B]^{\mathcal{A}}$ is block diagonal.

24.6. Let $\alpha \sim \beta$ indicate that the matrices α and β are congruent. Show that $\sim$ is an equivalence relation, that is show

    i. $\alpha \sim \alpha$ for all α (reflexive law)
    ii. $\alpha \sim \beta$ implies $\beta \sim \alpha$ (symmetric law)
    iii. $\alpha \sim \beta$ and $\beta \sim \gamma$ imply $\alpha \sim \gamma$ (transitive law)

24.7. Let $\alpha \in F^{n\times n}$ be the diagonal matrix $\alpha = \mathrm{diag}(a_1, a_2, \ldots, a_n)$ and let $b_1, b_2, \ldots, b_n$ be nonzero elements of F. Show that α is congruent to the matrix $\beta = \mathrm{diag}(a_1 b_1^2, a_2 b_2^2, \ldots, a_n b_n^2)$.

24.8. Let F be a field like the complex numbers in which every element has a square root and suppose $1+1 \neq 0$ in F. Using the above prove that every symmetric matrix is congruent to a diagonal matrix with diagonal entries 0 and 1 only.

24.9. Prove that every symmetric matrix over the real numbers is congruent to a diagonal matrix with diagonal entries 0, 1 and -1 only.

24.10. Let $\alpha \in F^{n\times n}$ be a skew-symmetric matrix and suppose α is congruent to a block matrix with k blocks $H = \begin{bmatrix} 0 & 1 \\ -1 & 0 \end{bmatrix}$. If $\mathcal{C}$ is any basis for $V = F^n$ and if $[B]^{\mathcal{C}} = \alpha$, show that $n - 2k = \dim_F \mathrm{rad}\, B$. Conclude that k is uniquely determined by α.

25. Inner Product Spaces

When we restrict our attention to certain special fields F, bilinear forms can become even more interesting. Here we will assume that $F = \mathbb{R}$ and start by posing a simple question about $\mathbb{R}^n$. Let $\alpha = (a_1, a_2, \ldots, a_n)$ and $\beta = (b_1, b_2, \ldots, b_n)$ be elements of $\mathbb{R}^n$. What is the angle θ between the line $\overline{0\alpha}$ joining 0 to α and the line $\overline{0\beta}$? This is actually a 2-dimensional problem since $0, \alpha$ and β are contained in the subspace $W = \langle \alpha, \beta \rangle$.

Let a denote the length of the line segment $\overline{0\alpha}$, b the length of the line segment $\overline{0\beta}$ and c the length of $\overline{\alpha\beta}$ as indicated on the diagram.

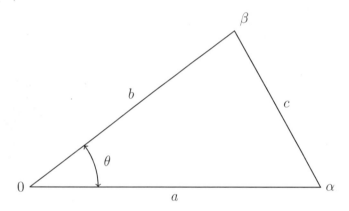

Then by the Pythagorean Theorem (and induction on n)

$$a^2 = a_1^2 + a_2^2 + \cdots + a_n^2$$
$$b^2 = b_1^2 + b_2^2 + \cdots + b_n^2$$

Furthermore, since c corresponds to the vector $\alpha - \beta$, we have

$$c^2 = (a_1 - b_1)^2 + (a_2 - b_2)^2 + \cdots + (a_n - b_n)^2$$

Now by the Law of Cosines

$$c^2 = a^2 + b^2 - 2ab\cos\theta$$

so

$$ab\cos\theta = \frac{1}{2}(a^2 + b^2 - c^2)$$
$$= a_1 b_1 + a_2 b_2 + \cdots + a_n b_n$$

and we have essentially found the angle θ.

Let us observe that the function of α and β given by the above right hand side is certainly a symmetric bilinear form which we now denote by $\alpha \bullet \beta$. In other words,

$$\alpha \bullet \beta = a_1 b_1 + a_2 b_2 + \cdots + a_n b_n$$

Moreover
$$\alpha \bullet \alpha = a_1^2 + a_2^2 + \cdots + a_n^2$$
and thus since $F = \mathbb{R}$ we have $\alpha \bullet \alpha \geq 0$ and $\alpha \bullet \alpha = 0$ if and only if $\alpha = 0$. In this way, $\mathbb{R}^n$ becomes an *inner product space*.

Formally an inner product space is a vector space V over R with an *inner product* or *dot product* $\alpha \bullet \beta$ defined on it satisfying

1. The inner product $\bullet$ maps $V \times V$ to $\mathbb{R}$ and is a symmetric bilinear form.
2. For all $\alpha \in V$, $\alpha \bullet \alpha \geq 0$ and $\alpha \bullet \alpha = 0$ if and only if $\alpha = 0$.

Let us consider some examples.

EXAMPLE 25.1. We have as above $V = \mathbb{R}^n$ and $\alpha \bullet \beta$ defined by
$$\alpha \bullet \alpha = a_1 b_1 + a_2 b_2 + \cdots + a_n b_n$$
where $\alpha = (a_1, a_2, \ldots, a_n)$ and $\beta = (b_1, b_2, \ldots, b_n)$.

EXAMPLE 25.2. Let V be a 3-dimensional vector space over $\mathbb{R}$ with basis $\mathcal{C} = \{\gamma_1, \gamma_2, \gamma_3\}$ and let $\alpha \bullet \beta = B(\alpha, \beta)$ where
$$[B]^{\mathcal{C}} = \begin{bmatrix} 2 & -1 & 0 \\ -1 & 2 & -2 \\ 0 & -2 & 4 \end{bmatrix}$$
Then $\bullet = B$ is certainly a symmetric bilinear form. Now suppose $\alpha \in V$ with $\alpha = a_1 \gamma_1 + a_2 \gamma_2 + a_3 \gamma_3$. Then
$$\alpha \bullet \alpha = B(\alpha, \alpha) = \sum_{i,j} a_i B(\gamma_i, \gamma_j) a_j$$
$$= 2a_1^2 - 2a_1 a_2 + 2a_2^2 - 4a_2 a_3 + 4a_3^2$$
$$= a_1^2 + (a_1 - a_2)^2 + (a_2 - 2a_3)^2$$
Thus we see that $\alpha \bullet \alpha \geq 0$ and that $\alpha \bullet \alpha = 0$ if and only if
$$a_1 = 0, \quad a_1 - a_2 = 0, \quad \text{and} \quad a_2 - 2a_3 = 0$$
and hence if and only if $\alpha = 0$.

EXAMPLE 25.3. Let $V = \mathbb{R}[x]$ and define
$$\alpha \bullet \beta = \int_{-1}^{1} \alpha(x) \beta(x) \, dx$$
If $\alpha = \sum_i a_i x^i$ and $\beta = \sum_i b_i x^i$, then
$$\alpha \bullet \beta = \sum_{i+j \text{ even}} \frac{2}{i+j+1} a_i b_j$$

Observe that $\alpha \bullet \alpha = \int_{-1}^{1} \alpha(x)^2 \, dx \geq 0$ and clearly $\alpha \bullet \alpha = 0$ if and only if $\alpha = 0$.

EXAMPLE 25.4. Let V be the subspace of all continuous real valued functions spanned by the trigonometric functions $\sin x, \sin 2x, \sin 3x, \ldots$ and $1, \cos x, \cos 2x, \cos 3x, \ldots$. For $\alpha, \beta \in V$ we define

$$\alpha \bullet \beta = \int_{-\pi}^{\pi} \alpha(x)\beta(x) \, dx$$

Since V consists of continuous functions of period 2π, we see easily that V is an inner product space.

Now let V be an inner product space. We say that α and β are *orthogonal* or perpendicular and write $\alpha \perp \beta$ if $\alpha \bullet \beta = 0$. We observe that in our original example

$$ab \cos \theta = \alpha \bullet \beta$$

and therefore if $\alpha, \beta \neq 0$, then $\alpha \bullet \beta = 0$ if and only if $\cos \theta = 0$. But $\cos \theta = 0$ mean that $\theta = \pm\pi/2$ and hence geometrically that $\overline{0\alpha}$ and $\overline{0\beta}$ are perpendicular lines.

If U and W are subspaces of V, we say that U and W are orthogonal if $\alpha \perp \beta$ for all $\alpha \in U$, $\beta \in W$. Similarly, $\sum_i W_i$ is an *orthogonal direct sum* if the sum is direct and if $W_i \perp W_j$ for all $i \neq j$.

LEMMA 25.1. *Let V be a finite dimensional inner product space over $\mathbb{R}$ and let W be a subspace of V.*

  i. *If $W \neq 0$, then W is nonisotropic.*
  ii. *W is nonsingular and hence $V = W \oplus W^{\perp}$.*
  iii. *$V = \sum_1^n W_i$ is an orthogonal direct sum of $n = \dim_{\mathbb{R}} V$ non-isotropic lines.*

PROOF. Most of this follows by virtue of the fact that the inner product is a symmetric bilinear form.

  (i) If $W \neq 0$ we can choose $0 \neq \alpha \in W$. By definition $\alpha \bullet \alpha > 0$ and hence W is nonisotropic.

  (ii) If $\alpha \in \operatorname{rad} W$ then we must have $\alpha \perp \alpha$ and thus $\alpha = 0$. Therefore $\operatorname{rad} W = 0$ and W is nonsingular so Theorem 22.2 implies that $V = W \oplus W^{\perp}$.

  (iii) By Theorem 23.1 we have $V = \sum_1^k W_i + U$, an orthogonal direct sum of the nonisotropic lines W_i and the isotropic subspace U. Since (i) implies that $U = 0$, the lemma is proved.   □

In our original example in $\mathbb{R}^n$, if $\alpha = (a_1, a_2, \ldots, a_n)$ then the length of the line segment $\overline{0\alpha}$ is given by

$$\sqrt{a_1^2 + a_2^2 + \cdots + a_n^2} = \sqrt{\alpha \bullet \alpha}$$

In general this is a parameter which in some sense measures the size of α. If V is any inner product space and if $\alpha \in V$, then we define the *norm* of α to be

$$\|\alpha\| = \sqrt{\alpha \bullet \alpha}$$

Observe that by definition we always have $\alpha \bullet \alpha \geq 0$ so it does make sense to take its square root in $\mathbb{R}$. If $\|\alpha\| = 1$ we say that α is a *normal vector*.

Let S be a subset of V. Then S is said to be *orthogonal* if for all $\alpha, \beta \in S$ with $\alpha \neq \beta$ we have $\alpha \perp \beta$. S is *orthonormal* if in addition for all $\alpha \in S$ we have $\|\alpha\| = 1$.

LEMMA 25.2. *Let $\alpha \in V$ and $a \in \mathbb{R}$. Then $\|a\alpha\| = |a|\cdot\|\alpha\|$. In particular, if $\alpha \neq 0$ and $a = 1/\|\alpha\|$, then $\|a\alpha\| = 1$.*

PROOF. We have

$$\|a\alpha\| = \sqrt{(a\alpha) \bullet (a\alpha)} = \sqrt{a^2(\alpha \bullet \alpha)} = |a|\cdot\|\alpha\|$$

In particular, if $\alpha \neq 0$ then $\|\alpha\| > 0$ and $\|a\alpha\| = 1$ when $a = 1/\|\alpha\|$. □

We now come to one of the important facts about finite dimensional inner product spaces.

THEOREM 25.1. *Let V be a finite dimensional inner product space. Then V has an orthonormal basis.*

PROOF. By part (iii) of Lemma 25.1, $V = \sum_1^n W_i$ is an orthogonal direct sum of nonisotropic lines $W_i - \langle \alpha_i \rangle$. It then follows easily that $\{\alpha_1, \alpha_2, \ldots, \alpha_n\}$ is an orthogonal basis for V. Finally, by the preceding lemma, for each α_i there exists $a_i = 1/\|\alpha_i\| \in \mathbb{R}$ with $\|a_i\alpha_i\| = 1$. Then $\{a_1\alpha_1, a_2\alpha_2, \ldots, a_n\alpha_n\}$ is certainly an orthonormal basis for V and the theorem is proved. □

We remark that the above result is decidedly false in general if V is allowed to be infinite dimensional. In the finite dimensional case there is actually a constructive procedure for finding an orthonormal basis known as the *Gram-Schmidt method*. It works as follows.

THEOREM 25.2. *Let V be an inner product space over $\mathbb{R}$ with basis $\{\alpha_1, \alpha_2, \ldots, \alpha_n\}$. We define vectors $\gamma_1, \gamma_2, \ldots, \gamma_n$ inductively by $\gamma_1 =$*

α_1 *and*

$$\gamma_{i+1} = \alpha_{i+1} - \sum_{k=1}^{i} \frac{(\alpha_{i+1} \bullet \gamma_k)}{(\gamma_k \bullet \gamma_k)} \gamma_k$$

Then, for all i

$$\mathcal{C}_i = \left\{ \frac{1}{\|\gamma_1\|} \gamma_1, \frac{1}{\|\gamma_2\|} \gamma_2, \ldots, \frac{1}{\|\gamma_i\|} \gamma_i \right\}$$

is an orthonormal basis for $\langle \alpha_1, \alpha_2, \ldots, \alpha_i \rangle$. *In particular* $\mathcal{C}_n$ *is an orthonormal basis for* V.

PROOF. We remark first that in the above formula for γ_{i+1} we have denominators of the form $(\gamma_k \bullet \gamma_k)$. Therefore we will have to show that each γ_k is nonzero.

The proof of the theorem proceeds by induction on i. When $i = 1$ we certainly have $\gamma_1 = \alpha_1 \neq 0$ so $\mathcal{C}_1 = \{ \frac{1}{\|\gamma_1\|} \gamma_1 \}$ is an orthonormal basis for $\langle \alpha_1 \rangle$ by Lemma 25.2.

Now assume that $\mathcal{C}_i$ is an orthonormal basis for $\langle \alpha_1, \alpha_2, \ldots, \alpha_i \rangle$. In particular each γ_k with $k \leq i$ is nonzero. Thus it makes sense to define

$$\gamma_{i+1} = \alpha_{i+1} - \sum_{k=1}^{i} \frac{(\alpha_{i+1} \bullet \gamma_k)}{(\gamma_k \bullet \gamma_k)} \gamma_k$$

Now observe that α_{i+1} can be solved from this equation and hence $\alpha_{i+1} \in \langle \gamma_1, \gamma_2, \ldots, \gamma_{i+1} \rangle$. But we also know, by induction, that $\alpha_k \in \langle \gamma_1, \gamma_2, \ldots, \gamma_i \rangle$ for all $k \leq i$. Hence

$$\langle \alpha_1, \alpha_2, \ldots, \alpha_{i+1} \rangle \subseteq \langle \gamma_1, \gamma_2, \ldots, \gamma_{i+1} \rangle$$

Now $\{ \alpha_1, \alpha_2, \ldots, \alpha_{i+1} \}$ is linearly independent, so the left hand subspace above has dimension $i + 1$. Since the right hand subspace is spanned by $i + 1$ vectors, it has dimension at most $i + 1$. The inclusion therefore becomes an equality and this implies that the $i + 1$ vectors $\{ \gamma_1, \gamma_2, \ldots, \gamma_{i+1} \}$ are linearly independent and in particular $\gamma_{i+1} \neq 0$. Moreover we now see that $\mathcal{C}_{i+1}$ is a basis for $\langle \alpha_1, \alpha_2, \ldots, \alpha_{i+1} \rangle$. We need only show that it is orthonormal.

By Lemma 25.2 we know that $\mathcal{C}_{i+1}$ consists of normal vectors and moreover by induction $\gamma_j \perp \gamma_k$ for all $j < k \leq i$. Thus we need only show that $\gamma_j \perp \gamma_{i+1}$ for all $j \leq i$. To this end, fix some $j \leq i$. Then by the definition of γ_{i+1} we have

$$\gamma_{i+1} \bullet \gamma_j = \left(\alpha_{i+1} - \sum_{k=1}^{i} \frac{(\alpha_{i+1} \bullet \gamma_k)}{(\gamma_k \bullet \gamma_k)}\gamma_k\right) \bullet \gamma_j$$

$$= \alpha_{i+1} \bullet \gamma_j - \sum_{k=1}^{i} \frac{(\alpha_{i+1} \bullet \gamma_k)}{(\gamma_k \bullet \gamma_k)}(\gamma_k \bullet \gamma_j)$$

Since $\gamma_k \bullet \gamma_j = 0$ in the above sum for all $k \neq j$, we have

$$\gamma_{i+1} \bullet \gamma_j = \alpha_{i+1} \bullet \gamma_j - \frac{(\alpha_{i+1} \bullet \gamma_j)}{(\gamma_j \bullet \gamma_j)}(\gamma_j \bullet \gamma_j) = 0$$

so $\gamma_{i+1} \perp \gamma_j$ and the theorem follows by induction. □

It is clear that the above result offers an elementary method for constructing an orthonormal basis.

Problems

25.1. Prove that an orthogonal set of nonzero vectors is always linearly independent.

25.2. We consider the inner product space $V = \mathbb{R}[x]$ with

$$\alpha \bullet \beta = \int_{-1}^{1} \alpha(x)\beta(x)\,dx$$

Let $D\colon R[x] \to R[x]$ denote the usual derivative d/dx and note that if $\alpha(x)$ is divisible by $\gamma(x)^k$ with $k \geq 1$ then $\alpha(x)D$ is divisible by $\gamma(x)^{k-1}$. For each integer n define $\alpha_n \in \mathbb{R}[x]$ by

$$\alpha_n(x) = (x^2 - 1)^n D^n$$

Show that α_n is a polynomial of degree n and that $\alpha_n \bullet \alpha_m = 0$ for all $m \neq n$. (Hint. If $m < n$ evaluate the integral $\alpha_m \bullet \alpha_n$ by parts $m + 1$ times, differentiating α_m and integrating α_n.)

25.3. Prove that $\{\alpha_0, \alpha_1, \alpha_2, \ldots\}$ is an orthogonal basis for the inner product space $V = \mathbb{R}[x]$ given above. The $\alpha_n(x)$ are suitable scalar multiples of the so-called *Legendre polynomials*.

25.4. Let m and n be integers. Evaluate the integrals

$$\int_{-\pi}^{\pi} \sin mx \cdot \sin nx\,dx, \quad \int_{-\pi}^{\pi} \sin mx \cdot \cos nx\,dx, \quad \int_{-\pi}^{\pi} \cos mx \cdot \cos nx\,dx$$

using appropriate trigonometric identities.

25.5. Let V be as in Example 25.4. Use the above to find an orthonormal basis for V.

25.6. Let V be a finite dimensional inner product space and let W be a subspace. Show that any orthonormal bases for W can be extended to an orthonormal bases for V.

25.7. Use the Gram-Schmidt method to find an orthonormal basis for the inner product space of Example 25.2.

25.8. Suppose $\{\alpha_1, \alpha_2, \ldots, \alpha_n\}$ is an orthonormal basis for V. What new orthonormal basis does the Gram-Schmidt method construct?

Let $\mathbb{Z}^+$ denote the set of positive integers and let V be the set of all bounded functions from $\mathbb{Z}^+$ to $\mathbb{R}$. That is $\alpha \in V$ if and only if $\alpha \colon \mathbb{Z}^+ \to \mathbb{R}$ and there exists a finite bound M (depending on α) with $\alpha(n) \le M$ for all $n \in \mathbb{Z}^+$. For $\alpha, \beta \in V$ define

$$\alpha \bullet \beta = \sum_{n=1}^{\infty} \frac{\alpha(n)\beta(n)}{2^n}$$

25.9. Prove that V is a vector space over $\mathbb{R}$ and in fact an inner product space.

25.10. Let W be the set of all elements $\alpha \in V$ such that $\alpha(n) = 0$ for all but finitely many $n \in \mathbb{Z}^+$. Show that W is a proper subspace of V and find $W^\perp$. Do you think V can have an orthonormal basis? See Problem 26.10.

26. Inequalities

Let V be a finite dimensional inner product space. Then we know that V has an orthonormal basis and we expect this basis to completely determine the inner product structure. Indeed we have

LEMMA 26.1. *Let V be an inner product space with orthonormal basis $\{\gamma_1, \gamma_2, \ldots, \gamma_n\}$. If $\alpha = \sum_i a_i \gamma_i$ and $\beta = \sum_i b_i \gamma_i$ are vectors in V, then $a_i = \alpha \bullet \gamma_i$ and $b_i = \beta \bullet \gamma_i$ so*

$$\alpha = \sum_i (\alpha \bullet \gamma_i)\gamma_i$$

$$\alpha \bullet \beta = \sum_i a_i b_i = \sum_i (\alpha \bullet \gamma_i)(\beta \bullet \gamma_i)$$

$$\|\alpha\|^2 = \sum_i a_i^2 = \sum_i (\alpha \bullet \gamma_i)^2$$

PROOF. By linearity we have

$$\alpha \bullet \beta = \left(\sum_i a_i \gamma_i\right) \bullet \left(\sum_j b_j \gamma_j\right)$$

$$= \sum_{i,j} a_i b_j \left(\gamma_i \bullet \gamma_j\right)$$

But $\gamma_i \bullet \gamma_j = 0$ for $i \neq j$ and $\gamma_i \bullet \gamma_i = 1$, so the above becomes

$$\alpha \bullet \beta = \sum_i a_i b_i$$

In particular, if $\beta = \gamma_i$ this yields $\alpha \bullet \gamma_i = a_i$ and similarly $\beta \bullet \gamma_i = b_i$. Thus $\alpha = \sum_i (\alpha \bullet \gamma_i)\gamma_i$ and

$$\alpha \bullet \beta = \sum_i (\alpha \bullet \gamma_i)(\beta \bullet \gamma_i)$$

Finally if $\beta = \alpha$ then

$$\|\alpha\|^2 = \alpha \bullet \alpha = \sum_i a_i^2 = \sum_i (\alpha \bullet \gamma_i)^2$$

and the lemma is proved.   □

Let us return for a moment to Example 25.1 and the comments at the beginning of the preceding section. Thus $V = \mathbb{R}^n$ and

$$ab \cos\theta = \alpha \bullet \beta$$

where $\alpha = (a_1, a_2, \ldots, a_n)$, $\beta = (b_1, b_2, \ldots, b_n)$, a is the length of $\overline{0\alpha}$, b is the length of $\overline{0\beta}$ and θ is the angle between the two line segments.

Clearly

$$a = \sqrt{a_1^2 + a_2^2 + \cdots + a_n^2} = \|\alpha\|$$

$$b = \sqrt{b_1^2 + b_2^2 + \cdots + b_n^2} = \|\beta\|$$

so

$$\cos\theta = \frac{\alpha \bullet \beta}{\|\alpha\|\cdot\|\beta\|}$$

Finally $|\cos\theta| \leq 1$ so we have shown here that

$$1 \geq \frac{\alpha \bullet \beta}{\|\alpha\|\cdot\|\beta\|}$$

While the above inequality is of interest, the above proof is certainly unsatisfying. It is based on certain geometric reasoning that just is not set on as firm a foundation as our algebraic reasoning. Admittedly, geometry can be axiomatized in such a way that the above proof will have validity in our situation, but most of us have not seen this formal approach to the subject. Thus it is best for us to look elsewhere for a proof of this inequality. In fact, we will offer three entirely different proofs of the result, the first of which is probably best since it avoids the use of bases.

The following is called the *Cauchy-Schwarz inequality*.

THEOREM 26.1. *Let V be an inner product space and let $\alpha, \beta \in V$. Then*

$$\|\alpha\|\cdot\|\beta\| \geq |\alpha \bullet \beta|$$

Moreover equality occurs if and only if one of α or β is a scalar multiple of the other.

PROOF. We observe that the theorem is trivially true if one of α or β is zero. Thus we can certainly assume that both α and β are nonzero. Moreover if α and β are scalar multiples of each other then we easily get equality here. Finally since α and β are contained in $\langle \alpha, \beta \rangle$, a finite dimensional subspace of V, we may clearly assume that V is finite dimensional.

FIRST PROOF. Let $x \in \mathbb{R}$. Then

$$0 \leq \|\alpha - x\beta\|^2 = (\alpha - x\beta) \bullet (\alpha - x\beta)$$
$$= \|\alpha\|^2 - 2x\,(\alpha \bullet \beta) + x^2\,\|\beta\|^2$$

What we have here is a parabola

$$y = \|\alpha\|^2 - 2x\,(\alpha \bullet \beta) + x^2\,\|\beta\|^2$$

in the (x, y)-plane that does not go below the x-axis. We obviously derive the most information from this by choosing x to be that real number that minimizes y, namely we take $x = (\alpha \bullet \beta)/\|\beta\|^2$. Then

$$0 \leq \|\alpha\|^2 - 2\frac{(\alpha \bullet \beta)}{\|\beta\|^2}(\alpha \bullet \beta) + \frac{(\alpha \bullet \beta)^2}{\|\beta\|^4}\|\beta\|^2$$

$$= \|\alpha\|^2 - \frac{(\alpha \bullet \beta)^2}{\|\beta\|^2}$$

This yields $(\alpha \bullet \beta)^2 \leq \|\alpha\|^2 \|\beta\|^2$ and the first part follows. Moreover we have equality if and only if, for this particular x, we have $\|\alpha - x\beta\|^2 = 0$ and hence if and only if $\alpha = x\beta$. □

SECOND PROOF. We may suppose that V is finite dimensional so that by Theorem 25.1, V has an orthonormal basis $\{\gamma_1, \gamma_2, \ldots, \gamma_n\}$. If $\alpha = \sum_i a_i \gamma_i$ and $\beta = \sum_i b_i \gamma_i$ then by Lemma 26.1

$$2\Big(\|\alpha\|^2 \|\beta\|^2 - (\alpha \bullet \beta)^2\Big)$$

$$= \Big(\sum_i a_i^2\Big)\Big(\sum_j b_j^2\Big) + \Big(\sum_i b_i^2\Big)\Big(\sum_j a_j^2\Big)$$

$$- 2\Big(\sum_i a_i b_i\Big)\Big(\sum_j a_j b_j\Big)$$

$$= \sum_{i,j}(a_i^2 b_j^2 + b_i^2 a_j^2 - 2a_i b_i a_j b_j)$$

$$= \sum_{i,j}(a_i b_j - b_i a_j)^2 \geq 0$$

Moreover if say $a_1 \neq 0$ then equality implies that $a_i b_1 - b_i a_1 = 0$ for all i so since $\beta \neq 0$ we have $b_1 \neq 0$ and then $a_i/a_1 = b_i/b_1$. Thus $(1/a_1)\alpha = (1/b_1)\beta$. □

THIRD PROOF. Again we may suppose that V is finite dimensional so we can extend $\{\beta\}$ to a basis $\{\beta_1, \beta_2, \ldots, \beta_n\}$ of V with $\beta_1 = \beta$. We then apply the Gram-Schmidt procedure to this basis and get an orthonormal basis $\{\gamma_1, \gamma_2, \ldots, \gamma_n\}$ of V with

$$\gamma_1 = \frac{1}{\|\beta_1\|}\beta_1 = \frac{1}{\|\beta\|}\beta$$

Then by Lemma 26.1

$$\|\alpha\|^2 = \sum_i (\alpha \bullet \gamma_i)^2 \geq (\alpha \bullet \gamma_1)^2$$

$$= \left(\alpha \bullet \frac{1}{\|\beta\|}\beta\right)^2 = \frac{1}{\|\beta\|^2}(\alpha \bullet \beta)^2$$

and the inequality is proved. Finally, if equality occurs then we must have $\|\alpha\|^2 = (\alpha \bullet \gamma_1)^2$ so $(\alpha \bullet \gamma_i) = 0$ for all $i > 1$ and $\alpha = (\alpha \bullet \gamma_1)\gamma_1$ is a scalar multiple of β.   □

This completes all three proofs of the theorem.   □

The following application is known as the *Triangle Inequality* for reasons that will be explained later.

THEOREM 26.2. *Let V be an inner product space and let $\alpha, \beta \in V$. Then*

$$\|\alpha + \beta\| \leq \|\alpha\| + \|\beta\|$$

PROOF. We have

$$\|\alpha + \beta\|^2 = (\alpha + \beta) \bullet (\alpha + \beta) = (\alpha \bullet \alpha) + 2(\alpha \bullet \beta) + (\beta \bullet \beta)$$

Since $\alpha \bullet \beta \leq \|\alpha\| \, \|\beta\|$ this yields

$$\|\alpha + \beta\|^2 \leq \|\alpha\|^2 + 2\|\alpha\| \, \|\beta\| + \|\beta\|^2 = (\|\alpha\| + \|\beta\|)^2$$

and thus, by taking square roots, we have $\|\alpha + \beta\| \leq \|\alpha\| + \|\beta\|$.   □

Suppose we return to our original example concerning $\mathbb{R}^n$ in this section and let $\alpha = (a_1, a_2, \ldots, a_n)$ and $\beta = (b_1, b_2, \ldots, b_n)$. Then the length of the line segment $\overline{\alpha\beta}$ is

$$\sqrt{(a_1 - b_1)^2 + (a_2 - b_2)^2 + \cdots + (a_n - b_n)^2} = \|\alpha - \beta\|$$

Now let γ be a third vector in $\mathbb{R}^n$ and form the triangle with α, β and γ as vertices. Since a straight line is the shortest distance between two points, it follows that the sum of the lengths of two sides of the triangle is greater than the length of the third. In particular we must have

$$\|\alpha - \gamma\| + \|\gamma - \beta\| \geq \|\alpha - \beta\|$$

Since $(\alpha - \gamma) + (\gamma - \beta) = \alpha - \beta$, this statement is an immediate consequence of the above Triangle Inequality.

Finally we consider an elementary formula that is known as the *Parallelogram Law.*

LEMMA 26.2. *Let V be an inner product space and let $\alpha, \beta \in V$. Then*

$$\|\alpha - \beta\|^2 + \|\alpha + \beta\|^2 = 2\|\alpha\|^2 + 2\|\beta\|^2$$

PROOF. We merely observe that

$$\|\alpha - \beta\|^2 + \|\alpha + \beta\|^2 = (\alpha - \beta) \bullet (\alpha - \beta) + (\alpha + \beta) \bullet (\alpha + \beta)$$

$$= \left(\|\alpha\|^2 - 2(\alpha \bullet \beta) + \|\beta\|^2 \right)$$

$$+ \left(\|\alpha\|^2 + 2(\alpha \bullet \beta) + \|\beta\|^2 \right)$$

$$= 2\|\alpha\|^2 + 2\|\beta\|^2$$

and the result follows. $\qquad\qquad\qquad\qquad\qquad\qquad\qquad\qquad\qquad\qquad\square$

Again consider $\mathbb{R}^n$ and nonzero vectors α and β. Then the quadrilateral drawn below with vertices $0, \alpha, \alpha + \beta$ and β is clearly a parallelogram. Thus there are two sides of length $\|\alpha\|$ and two of length $\|\beta\|$. On the other hand the diagonals have lengths $\|\alpha + \beta\|$ and $\|\alpha - \beta\|$. Thus the above lemma asserts that the sum of the squares of the lengths of the four sides of a parallelogram is equal to the sum of the squares of the lengths of the two diagonals.

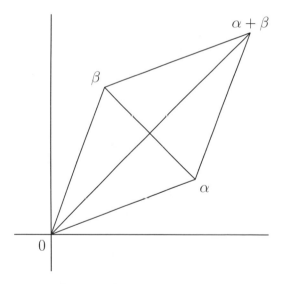

It is interesting to observe that this essentially trivial parallelogram law really characterizes the norm map. The following proof is not really algebraic. It uses limit properties of the real numbers.

THEOREM 26.3. *Let V be a vector space over $\mathbb{R}$ and let $\| \ \|: V \to \mathbb{R}$* *satisfy*

> *i.* $\|\alpha\| \geq 0$ *for all α and* $\|\alpha\| = 0$ *if and only if $\alpha = 0$.*
> *ii.* $\|r\alpha\| = |r| \cdot \|\alpha\|$ *for all $r \in \mathbb{R}$.*
> *iii.* $\|\alpha - \beta\|^2 + \|\alpha + \beta\|^2 = 2\|\alpha\|^2 + 2\|\beta\|^2$ *for all $\alpha, \beta \in V$.*

Then V is an inner product space with $\| \ \|$ as the associated norm.

PROOF. Using (iii) we define a function $V \times V \to \mathbb{R}$ by the two equivalent formulas

$$\alpha \bullet \beta = \frac{1}{2}\left(\|\alpha + \beta\|^2 - \|\alpha\|^2 - \|\beta\|^2\right)$$
$$= \frac{1}{2}\left(\|\alpha\|^2 + \|\beta\|^2 - \|\alpha - \beta\|^2\right)$$

and we show that this is an inner product. Certainly $\alpha \bullet \beta = \beta \bullet \alpha$ and

$$\alpha \bullet \alpha = \frac{1}{2}\left(\|\alpha\|^2 + \|\alpha\|^2 - \|0\|^2\right) = \|\alpha\|^2$$

so by (i) $\alpha \bullet \alpha \geq 0$ and $\alpha \bullet \alpha = 0$ if and only if $\alpha = 0$. It remains to show that $\alpha \bullet \beta$ is bilinear.

For convenience, let us rewrite (iii) as

$$\|\sigma - \tau\|^2 + \|\sigma + \tau\|^2 = 2\|\sigma\|^2 + 2\|\tau\|^2$$

and we apply this identity three times with the substitutions: (1) $\sigma = \alpha + \beta + \gamma$, $\tau = \alpha$; (2) $\sigma = \alpha + \beta$, $\tau = \alpha + \gamma$; (3) $\sigma = \beta$, $\tau = \gamma$. We obtain

(1) $2\|\alpha + \beta + \gamma\|^2 + 2\|\alpha\|^2 = \|2\alpha + \beta + \gamma\|^2 + \|\beta + \gamma\|^2$

(2) $\|2\alpha + \beta + \gamma\|^2 + \|\beta - \gamma\|^2 = 2\|\alpha + \beta\|^2 + 2\|\alpha + \gamma\|^2$

(3) $2\|\beta\|^2 + 2\|\gamma\|^2 = \|\beta + \gamma\|^2 + \|\beta - \gamma\|^2$

Adding these, cancelling the terms $\|2\alpha + \beta + \gamma\|^2$ and $\|\beta - \gamma\|^2$, and dividing by 2 yields easily

$$\|\alpha + \beta + \gamma\|^2 + \|\alpha\|^2 + \|\beta\|^2 + \|\gamma\|^2 \qquad\qquad (*)$$
$$= \|\alpha + \beta\|^2 + \|\alpha + \gamma\|^2 + \|\beta + \gamma\|^2$$

Now, by definition, we have

$$2\big(\alpha \bullet (\beta + \gamma)\big) - 2(\alpha \bullet \beta) - 2(\alpha \bullet \gamma)$$
$$= \left(\|\alpha + \beta + \gamma\|^2 - \|\alpha\|^2 - \|\beta + \gamma\|^2\right)$$
$$- \left(\|\alpha + \beta\|^2 - \|\alpha\|^2 - \|\beta\|^2\right)$$
$$- \left(\|\alpha + \gamma\|^2 - \|\alpha\|^2 - \|\gamma\|^2\right)$$
$$= \left(\|\alpha + \beta + \gamma\|^2 + \|\alpha\|^2 + \|\beta\|^2 + \|\gamma\|^2\right)$$
$$- \left(\|\alpha + \beta\|^2 + \|\alpha + \gamma\|^2 + \|\beta + \gamma\|^2\right)$$
$$= 0$$

by equation $(*)$ above. Thus

$$\alpha \bullet (\beta + \gamma) = \alpha \bullet \beta + \alpha \bullet \gamma$$

and by symmetry we also have

$$(\beta + \gamma) \bullet \alpha = \beta \bullet \alpha + \gamma \bullet \alpha$$

Let m, n be positive integers. Then induction yields $\alpha \bullet (n\beta) = n(\alpha \bullet \beta)$ so

$$m\left(\alpha \bullet \frac{n}{m}\beta\right) = \alpha \bullet n\beta = n(\alpha \bullet \beta)$$

and

$$\alpha \bullet \frac{n}{m}\beta = \frac{n}{m}(\alpha \bullet \beta)$$

Moreover, by definition and (ii) we have

$$\alpha \bullet (-\beta) = \frac{1}{2}\left(\|\alpha\|^2 + \|-\beta\|^2 - \|\alpha + \beta\|^2\right)$$

$$= -\frac{1}{2}\left(\|\alpha + \beta\|^2 - \|\alpha\|^2 - \|\beta\|^2\right) = -(\alpha \bullet \beta)$$

and this shows that for all $b \in \mathbb{Q}$, $\alpha \bullet b\beta = b(\alpha \bullet \beta)$. Moreover, for all $a \in \mathbb{Q}$, symmetry yields $a\alpha \bullet \beta = a(\alpha \bullet \beta)$.

Let $r \in \mathbb{R}$ with $r \neq 0$ and let $a \in \mathbb{Q}$ with $a > 0$. Then by (ii)

$$r\alpha \bullet a\beta = \frac{1}{2}\left(\|r\alpha\|^2 + \|a\beta\|^2 - \|r\alpha - a\beta\|^2\right)$$

$$\leq \frac{1}{2}\left(\|r\alpha\|^2 + \|a\beta\|^2\right)$$

$$= \frac{1}{2}r^2\|\alpha\|^2 + \frac{1}{2}a^2\|\beta\|^2$$

Since $r\alpha \bullet a\beta = a(r\alpha \bullet \beta)$ and $a > 0$ this yields

$$r\alpha \bullet \beta \leq \frac{r^2}{2a}\|\alpha\|^2 + \frac{a}{2}\|\beta\|^2$$

If we now choose a so that $|r|/2 < a < 2|r|$ the above easily becomes

$$r\alpha \bullet \beta \leq |r|\left(\|\alpha\|^2 + \|\beta\|^2\right)$$

Moreover replacing β by $-\beta$ and using $\|-\beta\| = \|\beta\|$ we get

$$-(r\alpha \bullet \beta) = r\alpha \bullet (-\beta) \leq |r|\left(\|\alpha\|^2 + \|\beta\|^2\right)$$

and thus

$$|(r\alpha \bullet \beta)| \leq |r|\left(\|\alpha\|^2 + \|\beta\|^2\right) \qquad (**)$$

Finally fix $\alpha, \beta \in V$ and $s \in \mathbb{R}$. We choose $c \in \mathbb{Q}$ with c "close to but not equal to" s and obtain

$$
\begin{aligned}
|s\alpha \bullet \beta - s(\alpha \bullet \beta)| &= \left| \big(s\alpha \bullet \beta - s(\alpha \bullet \beta) \big) - \big(c\alpha \bullet \beta - c(\alpha \bullet \beta) \big) \right| \\
&= |(s-c)\alpha \bullet \beta - (s-c)(\alpha \bullet \beta)| \\
&\leq |(s-c)\alpha \bullet \beta| + |s-c||\alpha \bullet \beta| \\
&\leq |s-c| \Big(\|\alpha\|^2 + \|\beta\|^2 + |\alpha \bullet \beta| \Big)
\end{aligned}
$$

by equation (∗∗) with $r = s - c$. Now letting $c \in \mathbb{Q}$ approach s, we observe that the above right hand side approaches 0 and we conclude that

$$(s\alpha) \bullet \beta = s(\alpha \bullet \beta)$$

By symmetry

$$\alpha \bullet (s\beta) = s(\alpha \bullet \beta)$$

so $\bullet$ is a bilinear form and the result follows. $\square$

Problems

26.1. Prove geometrically that the sum of the squares of the lengths of the four sides of a parallelogram is equal to the sum of the squares of the lengths of the two diagonals. (Hint. An obvious approach is to try two applications of the Law of Cosines.)

Let V be an inner product space.

26.2. Show that equality occurs in the Cauchy-Schwarz inequality if $\alpha = a\beta$ or $\beta = b\alpha$ for some $a, b \in \mathbb{R}$.

26.3. Prove that $\big| \|\alpha\| - \|\beta\| \big| \leq \|\alpha - \beta\|$ for all $\alpha, \beta \in V$.

A linear transformation $T\colon V \to V$ is said to be *unitary* if $\|\alpha T\| = \|\alpha\|$ for all $\alpha \in V$.

26.4. Let $T\colon V \to V$ be a linear transformation. Show that T is unitary if and only if $(\alpha T) \bullet (\beta T) = \alpha \bullet \beta$ for all $\alpha, \beta \in V$.

26.5. Let $\mathcal{C} = \{\gamma_1, \gamma_2, \ldots, \gamma_n\}$ be an orthonormal basis for V. Prove that $T\colon V \to V$ is unitary if and only if $(\mathcal{C})T = \{\gamma_1 T, \gamma_2 T, \ldots, \gamma_n T\}$ is also an orthonormal basis.

26.6. Let $\mathcal{C}$ be as above. Show that T is unitary if and only if $c[T]c \cdot c[T]c^{\mathsf{T}} = I_n$ where ^{T} is the matrix transpose map.

Let V denote the vector space of all bounded real valued functions $\alpha \colon \mathbb{Z}^+ \to \mathbb{R}$ as defined immediately before Problem 25.9 and recall that

$$\alpha \bullet \beta = \sum_{n=1}^{\infty} \frac{\alpha(n)\beta(n)}{2^n}$$

Then we know that V is an inner product space. Suppose by way of contradiction that V has an orthonormal basis $\mathcal{C}$.

26.7. Show that V is infinite dimensional and therefore that $\mathcal{C}$ has a countably infinite subset $\{\gamma_1, \gamma_2, \ldots, \gamma_i, \ldots\}$.

26.8. For each γ_i as above, let $c_i > 0$ denote a finite upper bound for $\{|\gamma_i(n)| \mid n \in \mathbb{Z}^+\}$. Now define $\alpha \colon \mathbb{Z}^+ \to \mathbb{R}$ by

$$\alpha(n) = \sum_{i=1}^{\infty} \frac{\gamma_i(n)}{2^i c_i}$$

(Note that we are not adding infinitely many functions since this is not defined on V, but rather we are adding infinitely many real function values.) Prove that $\alpha \in V$.

26.9. For all i show that

$$\alpha \bullet \gamma_i = \frac{1}{2^i c_i} \neq 0$$

(Hint. $\alpha \bullet \gamma_i$ is a double sum and can be computed by interchanging the order of summation and using the values of the various $\gamma_j \bullet \gamma_i$.)

26.10. Use the preceding problem and the ideas of Lemma 26.1 to deduce that when α is written in terms of the basis $\mathcal{C}$ that all γ_i must occur. Conclude that V does not have an orthonormal basis.

27. Real Symmetric Matrices

Here we study real symmetric matrices. As in the preceding sections, it will be easier to study certain vector spaces and then translate the results into matrix language. We will consider two different properties of the matrix, namely its behavior as a bilinear form and then as a linear transformation.

Let B be a bilinear form on the real vector space V and let W be a subspace. We say that W is a *positive semidefinite* subspace if $B(\beta, \beta) \geq 0$ for all $\beta \in W$, We say that W is *positive definite* if in addition $B(\beta, \beta) = 0$ only when $\beta = 0$. Analogously, we say that W is *negative semidefinite* if $B(\beta, \beta) \leq 0$ for all $\beta \in W$ and W is *negative definite* if in addition $B(\beta, \beta) = 0$ only when $\beta = 0$. The following is known as *Sylvester's Law of Inertia*.

THEOREM 27.1. *Let* $B \colon V \times V \to \mathbb{R}$ *be a symmetric bilinear form on the* n*-dimensional real vector space* V. *Then* V *has a basis* $\mathcal{C} = \{\gamma_1, \gamma_2, \ldots, \gamma_n\}$ *with*

$$[B]^{\mathcal{C}} = \operatorname{diag}(1, 1, \ldots, 1, -1, -1, \ldots, -1, 0, 0, \ldots, 0)$$

Moreover the number of 1*'s,* -1*'s and* 0*'s that occur on the diagonal are uniquely determined by* B.

PROOF. By Theorem 23.1, we can write

$$V = W_1 + W_2 + \cdots + W_k + U$$

a perpendicular direct sum of the nonisotropic lines W_i and the isotropic subspace U. Let $W_i = \langle \alpha_i \rangle$ and let $\{\alpha_{k+1}, \ldots, \alpha_n\}$ be a basis for U. It follows that $\mathcal{A} = \{\alpha_1, \alpha_2, \ldots, \alpha_k, \ldots, \alpha_n\}$ is a basis for V and that $[B]^{\mathcal{A}}$ is diagonal with diagonal entries $B(\alpha_i, \alpha_i) = a_i$. Furthermore, if $0 \neq c_i \in \mathbb{R}$ and if $\gamma_i = c_i \alpha_i$, then $\mathcal{C} = \{\gamma_1, \gamma_2, \ldots, \gamma_n\}$ is also a basis for V. Clearly

$$[B]^{\mathcal{C}} = \operatorname{diag}\big(B(\gamma_1, \gamma_1), B(\gamma_2, \gamma_2), \ldots, B(\gamma_n, \gamma_n)\big)$$
$$= \operatorname{diag}\big(c_1^2 a_1, c_2^2 a_2, \ldots, c_n^2 a_n\big)$$

Now we can certainly choose $c_i \neq 0$ so that $c_i^2 a_i = 1, -1$ or 0 accordingly as $a_i > 0$, $a_i < 0$ or $a_i = 0$ and we do so. Finally, we order the vectors in $\mathcal{C}$ so that the $+1$ entries come first, then the -1's and lastly the 0's. In this way

$$[B]^{\mathcal{C}} = \operatorname{diag}(\underbrace{1, 1, \ldots, 1}_{r}, \underbrace{-1, -1, \ldots, -1}_{s}, \underbrace{0, 0, \ldots, 0}_{t})$$

Let r, s and t be as above. We show that these parameters are uniquely determined by B. To this end, let us first make a notational change so that the orthogonal basis $\mathcal{C}$ is written as

$$\mathcal{C} = \{\alpha_1, \alpha_2, \ldots, \alpha_r, \beta_1, \beta_2, \ldots, \beta_s, \gamma_1, \gamma_2, \ldots, \gamma_t\}$$

with $B(\alpha_i, \alpha_i) = 1$, $B(\beta_i, \beta_i) = -1$ and $B(\gamma_i, \gamma_i) = 0$. Set

$$W^+ = \langle \alpha_1, \alpha_2, \ldots, \alpha_r \rangle, \quad W^- = \langle \beta_1, \beta_2, \ldots, \beta_s, \gamma_1, \gamma_2 \ldots, \gamma_t \rangle$$

Suppose $\delta \in W^+$. Then $\delta = \sum_i d_i \alpha_i$ so

$$B(\delta, \delta) = \sum_i B(d_i \alpha_i, d_i \alpha_i) = \sum_i d_i^2$$

Thus $B(\delta, \delta) \geq 0$ and $B(\delta, \delta) = 0$ if and only if $\delta = 0$. In other words, W^+ is a positive definite subspace and hence V has a positive definite subspace of dimension r.

Now let $\delta \in W^-$. Then $\delta = \sum_i d_i \beta_i + \sum_j e_j \gamma_j$ so

$$B(\delta, \delta) = \sum_i B(d_i \beta_i, d_i \beta_i) + \sum_j B(e_j \gamma_j, e_j \gamma_j) = -\sum_i d_i^2 \leq 0$$

and thus W^- is a negative semidefinite subspace of dimension equal to $s + t = n - r$. Finally let W be any positive definite subspace of V. If $\delta \in W \cap W^-$, then $B(\delta, \delta) \geq 0$ since $\delta \in W$ and $B(\delta, \delta) \leq 0$ since $\delta \in W^-$. Thus $B(\delta, \delta) = 0$ and since W is positive definite we have $\delta = 0$. Therefore $W \cap W^- = 0$ so

$$\dim_{\mathbb{R}} W + \dim_{\mathbb{R}} W^- \leq \dim_{\mathbb{R}} V = n$$

and hence $\dim_{\mathbb{R}} W \leq r$ since $\dim_{\mathbb{R}} W^- = n - r$. It follows that r is the largest dimension of a positive definite subspace of V and therefore r is uniquely determined by B.

We obtain s in a similar manner by reversing the roles of the α's and the β's. We set

$$W^- = \langle \beta_1, \beta_2, \ldots, \beta_s \rangle, \quad W^+ = \langle \alpha_1, \alpha_2, \ldots, \alpha_r, \gamma_1, \gamma_2 \ldots, \gamma_t \rangle$$

and as above we conclude that s is the largest dimension of a negative definite subspace of V. Thus s is uniquely determined by B and since $t = n - r - s$, the theorem is proved. $\square$

We can now translate the above into a result on matrices.

COROLLARY 27.1. *Let $\alpha \in \mathbb{R}^{n \times n}$ be a symmetric matrix. Then α is congruent to a diagonal matrix with diagonal entries 1, -1 and 0 only. Moreover the number of 1's, -1's and 0's is uniquely determined by α.*

PROOF. Choose some n-dimensional real vector space V, say $V = \mathbb{R}^n$, and fix a basis $\mathcal{A}$ of V. Then, by Theorems 24.1 and 24.2, there exists a unique bilinear form $B: V \times V \to \mathbb{R}$ with $[B]^{\mathcal{A}} = \alpha$. By the preceding theorem, there exists a basis $\mathcal{C}$ of V such that $[B]^{\mathcal{C}}$ is diagonal with diagonal entries 1, -1 and 0 and with the number of these entries determined by B and hence by α. Since $\alpha = [B]^{\mathcal{A}}$ is congruent to $[B]^{\mathcal{C}}$ by Lemma 24.1, we see that α is congruent to an appropriate diagonal matrix.

Conversely assume that α is congruent to a suitable diagonal matrix γ and let $V = \mathbb{R}^n$ have basis $\mathcal{A}$. Then B can be chosen so that $\alpha = [B]^{\mathcal{A}}$ and B depends only upon α. Since α is congruent to γ, Lemma 24.1 implies that $[B]^{\mathcal{C}} = \gamma$ for some basis $\mathcal{C}$ of V. The preceding theorem now implies that the number of diagonal entries of γ equal to 1, -1 or 0 is uniquely determined by B and hence by α. $\qquad\square$

We now move on to consider linear transformation properties of real symmetric matrices. Again we start by studying vector spaces and linear transformations. Let V be an inner product space. A linear transformation $T\colon V \to V$ is said to be *symmetric* if for all $\alpha, \beta \in V$ we have

$$\alpha T \bullet \beta = \alpha \bullet \beta T$$

Recall that the complex numbers are algebraically closed, so every nonconstant polynomial in $\mathbb{C}[x]$ has a root in $\mathbb{C}$. It follows that every nonzero polynomial in $\mathbb{C}[x]$ can be factored into a product of linear factors. For the real numbers we have

LEMMA 27.1. *Every nonzero polynomial in $\mathbb{R}[x]$ can be factored into a product of linear and quadratic factors.*

PROOF. We proceed by induction on the degree of the polynomial f, the result being trivial for $\deg f = 0$ or 1. Now let $f(x) \in \mathbb{R}[x]$ be the nonzero polynomial

$$f(x) = \sum_{i=0}^{n} a_i x^i$$

of degree $n > 1$ and assume that the result holds for all polynomials of smaller degree. We embed $\mathbb{R}$ in the complex numbers $\mathbb{C}$ and then $\mathbb{R}[x] \subseteq \mathbb{C}[x]$.

Let $\bar{}\colon \mathbb{C} \to \mathbb{C}$ denote complex conjugation. Since $\mathbb{C}$ is algebraically closed, we can let $b \in \mathbb{C}$ be a root of f. Then since the coefficients a_i are real, we have

$$0 = \bar{0} = \overline{\sum a_i b^i} = \sum \bar{a}_i \bar{b}^i$$
$$= \sum a_i \bar{b}^i = f(\bar{b})$$

and $\bar{b}$ is also a root of f. By Lemma 18.4 we have $f(x) = (x - b)g(x)$ for some $g(x) \in \mathbb{C}[x]$. Indeed, if b is real, then that lemma implies that $g(x) \in \mathbb{R}[x]$, and the result follows by induction since $\deg g < \deg f$.

If b is not real, then $\bar{b} \neq b$ is also a root of f, so $\bar{b}$ is a root of g. Hence $g(x) = (x - \bar{b})h(x)$ and $f(x) = (x - b)(x - \bar{b})h(x)$. Finally, we see that

$$(x - b)(x - \bar{b}) = x^2 - (b + \bar{b})x + b\bar{b}$$

is a real quadratic polynomial. Furthermore by long division, $h(x) \in \mathbb{R}[x]$ and the result again follows by induction since $\deg h < \deg f$. $\square$

With this lemma, we can obtain

LEMMA 27.2. *Let V be a finite dimensional inner product space and let $T: V \to V$ be a symmetric linear transformation. Then T has an eigenvalue in $\mathbb{R}$.*

PROOF. By the Cayley-Hamilton Theorem, there exists a nonzero polynomial $f(x) \in \mathbb{R}[x]$ of minimal degree with $f(T) = 0$. Clearly $\deg f \geq 1$, and by the previous lemma, $f(x) = g(x)h(x)$ where $h(x)$ is monic and either linear or quadratic. Now $\deg g < \deg f$, so by the minimal nature of $\deg f$ we have $g(T) \neq 0$. Hence there exists $\gamma \in V$ with $\alpha = \gamma g(T) \neq 0$. Lemma 19.2 now implies that

$$\alpha h(T) = \gamma g(T)h(T) = \gamma f(T) = 0$$

If $h(x) = x - a$ is linear then

$$0 = \alpha h(T) = aT - a\alpha$$

and $a \in \mathbb{R}$ is an eigenvalue for T.

Suppose now that $h(x) = x^2 + 2ax + b$ is quadratic. Then

$$0 = \alpha h(T) = \alpha T^2 + 2a\alpha T + b\alpha$$

and hence

$$0 = (\alpha T^2 + 2a(\alpha T) + b\alpha) \bullet \alpha$$
$$= (\alpha T^2 \bullet \alpha) + 2a(\alpha T \bullet \alpha) + b(\alpha \bullet \alpha)$$

Note that T is symmetric, so $\alpha T^2 \bullet \alpha = \alpha T \bullet \alpha T = \|\alpha T\|^2$ and the above yields

$$0 = \|\alpha T\|^2 + 2a(\alpha T \bullet \alpha) + b\|\alpha\|^2 \qquad (*)$$

On the other hand

$$0 \leq \|\alpha T + a\alpha\|^2 = (\alpha T + a\alpha) \bullet (\alpha T + a\alpha)$$
$$= \|\alpha T\|^2 + 2a(\alpha T \bullet \alpha) + a^2\|\alpha\|^2$$

and subtracting $(*)$ from the above yields

$$0 \le (a^2 - b)\, \|\alpha\|^2$$

Since $\alpha \ne 0$ we have $\|\alpha\|^2 > 0$ and we conclude that $a^2 - b \ge 0$. Therefore the numbers $c_1 = -a + \sqrt{a^2 - b}$ and $c_2 = -a - \sqrt{a^2 - b}$ are real and these are the roots of $h(x)$ so $h(x) = (x - c_1)(x - c_2)$. This shows that $f(x)$ has a linear factor and as we have seen this implies that T has a real eigenvalue. $\qquad \square$

The existence of this one eigenvalue allows us to deduce a good deal of information about T. We have

THEOREM 27.2. *Let V be a finite dimensional inner product space and let $T\colon V \to V$ be a symmetric linear transformation. Then V has an orthonormal basis consisting of eigenvectors of T.*

PROOF. We proceed by induction on $\dim_{\mathbb{R}} V$, the result being trivial if $\dim_{\mathbb{R}} V = 1$. Let $\dim_{\mathbb{R}} V = n > 1$ and suppose that the result is true for all smaller dimensional spaces. Let $T\colon V \to V$ be symmetric. By the preceding lemma, T has an eigenvalue $a \in \mathbb{R}$ with corresponding eigenvector α. Since $b\alpha$ is also an eigenvector for any $b \in \mathbb{R}$ with $b \ne 0$, we can certainly assume that $\|\alpha\| = 1$. If $W = \langle \alpha \rangle^{\perp}$ then by Lemma 25.1 we have $V = \langle \alpha \rangle \oplus W$, an orthogonal direct sum. Clearly $\dim_{\mathbb{R}} W = n - 1$ and W is an inner product space.

We show now that $(W)T \subseteq W$. To this end, let $\beta \in W$. Since T is symmetric and $\beta \in \langle \alpha \rangle^{\perp}$ we have

$$\alpha \bullet \beta T = \alpha T \bullet \beta = (a\alpha) \bullet \beta = 0$$

Hence $\beta T \subseteq \langle \alpha \rangle^{\perp} = W$. Thus $T_W\colon W \to W$ and clearly T_W is symmetric. By induction W has an orthonormal basis $\{\alpha_2, \alpha_3, \dots, \alpha_n\}$ consisting of eigenvectors for T_W and hence for T. Setting $\alpha_1 = \alpha$ it follows that $\{\alpha_1, \alpha_2, \dots, \alpha_n\}$ is an orthonormal basis for V consisting of eigenvectors for T and the result follows. $\qquad \square$

Finally we have

COROLLARY 27.2. *Let $\alpha \in \mathbb{R}^{n \times n}$ be a symmetric matrix. Then α is similar to a diagonal matrix.*

PROOF. Let V be some n-dimensional inner product space, for example we could take $V = \mathbb{R}^n$ with the usual dot product. If $\mathcal{C} = \{\gamma_1, \gamma_2, \dots, \gamma_n\}$ is an orthonormal basis for V, then we can find a linear transformation $T\colon V \to V$ such that $_\mathcal{C}[T]_\mathcal{C} = \alpha$. We show now that T is symmetric.

Write $\alpha = [a_{ij}]$ and let $\beta = \sum_i b_i \gamma_i$ and $\delta = \sum_j d_j \gamma_j$ be vectors in V. Then

$$\beta T = \sum_{i,j} b_i a_{ij} \gamma_j$$

$$\delta T = \sum_{i,j} d_j a_{ji} \gamma_i$$

so since $\mathcal{C}$ is orthonormal we have

$$\beta T \bullet \delta = \left(\sum_{i,j} b_i a_{ij} \gamma_j \right) \bullet \left(\sum_j d_j \gamma_j \right)$$

$$= \sum_{i,j} b_i a_{ij} d_j$$

and

$$\beta \bullet \delta T = \left(\sum_i b_i \gamma_i \right) \bullet \left(\sum_{i,j} d_j a_{ji} \gamma_i \right)$$

$$= \sum_{i,j} b_i a_{ji} d_j$$

But α is symmetric, so $a_{ij} = a_{ji}$ and thus $\beta T \bullet \delta = \beta \bullet \delta T$ as required.

Finally by the previous theorem, V has an orthonormal basis $\mathcal{A}$ consisting of eigenvectors of T and hence $_{\mathcal{A}}[T]_{\mathcal{A}}$ is diagonal. Since $\alpha = {}_{\mathcal{C}}[T]_{\mathcal{C}}$ is similar to $_{\mathcal{A}}[T]_{\mathcal{A}}$, the proof is complete.   □

Problems

27.1. Let $\alpha \in \mathbb{R}^{n \times n}$. Show that α is similar to a diagonal matrix if and only if it is similar to a symmetric matrix.

27.2. Let V be an inner product space with orthonormal basis $\mathcal{C} = \{\gamma_1, \gamma_2, \ldots, \gamma_n\}$ and let $\alpha \in \mathbb{R}^{n \times n}$. Define the linear transformations T and T^{T} by

$$_{\mathcal{C}}[T]_{\mathcal{C}} = \alpha, \qquad _{\mathcal{C}}[T^{\mathsf{T}}]_{\mathcal{C}} = \alpha^{\mathsf{T}}$$

Prove that $\beta T \bullet \delta = \beta \bullet \delta T^{\mathsf{T}}$ for all vectors $\beta, \delta \in V$.

Let V be an. n-dimensional real vector space and let $B \colon V \times V \to \mathbb{R}$ be a symmetric bilinear form. Suppose there exists a basis $\mathcal{C}$ of V with

$$[B]^{\mathcal{C}} = \operatorname{diag}(\underbrace{1, 1, \ldots, 1}_{r}, \underbrace{-1, -1, \ldots, -1}_{s}, \underbrace{0, 0, \ldots, 0}_{t})$$

27.3. Let U be an isotropic subspace of V. Prove that $\dim_{\mathbb{R}} U \le n - r$ and $n - s$.

27.4. Show that there exists an isotropic subspace U of dimension $n - \max\{r, s\} = t + \min\{r, s\}$.

Let V be an n-dimensional real vector space having the basis $\mathcal{C} = \{\gamma_1, \gamma_2, \ldots, \gamma_n\}$ and let $B \colon V \times V \to \mathbb{R}$ be a symmetric bilinear form. Suppose α is the real symmetric matrix given by $[B]^{\mathcal{C}} = \alpha$. We obtain a condition on α equivalent to B being an inner product.

27.5. Suppose B is an inner product. Deduce that α is congruent to the identity matrix and hence that $\det \alpha > 0$.

27.6. For each i, let $V_i \subseteq V$ be given by $V_i = \langle \gamma_1, \gamma_2, \ldots, \gamma_i \rangle$ so that $\mathcal{C}_i = \{\gamma_1, \gamma_2, \ldots, \gamma_i\}$ is a basis for V_i. Show that the matrix $[B_{V_i}]^{\mathcal{C}_i}$ is equal to α_i, the square submatrix of α formed by the first i rows and first i columns of α.

27.7. If B is an inner product, then obviously so is B_W for all subspaces W of V. Using this, conclude that $\det \alpha_i > 0$ for all subscripts $i = 1, 2, \ldots, n$.

27.8. Let $W = V_{n-1}$ and suppose that B_W is an inner product. Then certainly $V = W \oplus W^\perp$ with $W^\perp = \langle \beta \rangle$ for some nonzero vector β. Let $\mathcal{B}$ be the basis of V given by $\mathcal{B} = \{\gamma_1, \gamma_2, \ldots, \gamma_{n-1}, \beta\}$ and find the matrix $[B]^{\mathcal{B}}$. Show that B is an inner product if and only if $B(\beta, \beta) > 0$ and hence if and only if $\det[B]^{\mathcal{B}} > 0$.

27.9. Given the same situation as in Problem 27.8, use the fact that $[B]^{\mathcal{B}}$ is congruent to α to deduce that B is an inner product if and only if $\det \alpha > 0$.

27.10. Prove by induction on n that B is an inner product if and only if $\det \alpha_i > 0$ for $i = 1, 2, \ldots, n$.

28. Complex Analogs

When dealing with complex vector spaces, somewhat different bilinear forms and inner products occur. They depend upon the fact that $\mathbb{C}$ has a conjugation map $^-\colon \mathbb{C} \to \mathbb{C}$ such that $a\bar{a} = |a|^2 \geq 0$ for all $a \in \mathbb{C}$. Most of the theory follows in a completely analogous manner and therefore we will confine our proofs to the few special situations in which differences occur.

Let V be a complex vector space. A map $B\colon V \times V \to \mathbb{C}$ is said to be a *Hermitian bilinear form* if for all $\alpha, \beta, \gamma \in V$ and $a, b \in \mathbb{C}$ we have

H1.
$$B(\alpha + \beta, \gamma) = B(\alpha, \gamma) + B(\beta, \gamma)$$
$$B(\gamma, \alpha + \beta) = B(\gamma, \alpha) + B(\gamma, \beta)$$

H2.
$$B(a\alpha, \beta) = aB(\alpha, \beta)$$
$$B(\alpha, b\beta) = \bar{b}B(\alpha, \beta)$$

Thus we observe that B differs from the usual bilinear form just in the last equation where the scalar b factors out as $\bar{b}$, its complex conjugate. We consider some examples.

EXAMPLE 28.1. Let $V = \mathbb{C}^n$. Then we define a dot product $\alpha \bullet \beta = B(\alpha, \beta)$ by

$$\alpha \bullet \beta = B(\alpha, \beta) = \sum_i a_i \bar{b}_i$$

where $\alpha = (a_1, a_2, \ldots, a_n)$ and $\beta = (b_1, b_2, \ldots, b_n)$. It is easy to see that this is a Hermitian bilinear form. It has a number of additional properties we will find of interest later.

EXAMPLE 28.2. More generally, let V be an n-dimensional $\mathbb{C}$-vector space with basis $\{\gamma_1, \gamma_2, \ldots, \gamma_n\}$ and let $[c_{ij}] \in \mathbb{C}^{n \times n}$. Then $B\colon V \times V \to \mathbb{C}$ given by

$$B\left(\sum_i a_i \gamma_i, \sum_j b_j \gamma_j\right) = \sum_{i,j} a_i c_{ij} \bar{b}_j$$

is a Hermitian bilinear form.

In other words, we can easily construct such forms by taking an ordinary bilinear form and somehow conjugating the second factor.

THEOREM 28.1. *Let V be an n-dimensional $\mathbb{C}$-vector space with basis $\mathcal{C} = \{\gamma_1, \gamma_2, \ldots, \gamma_n\}$. Then the map*

$$B \mapsto [B]^{\mathcal{C}} = [B(\gamma_i, \gamma_j)]$$

yields a one-to-one correspondence between Hermitian bilinear forms B and all $n \times n$ matrices over $\mathbb{C}$.

Let us observe an interesting difference here.

LEMMA 28.1. *Suppose* $B : V \times V \to \mathbb{C}$ *is a nonzero Hermitian bilinear form. Then there exists* $\alpha \in V$ *with* $B(\alpha, \alpha) \neq 0$.

PROOF. Since B is nonzero there exists $\beta, \gamma \in V$ with $B(\beta, \gamma) \neq 0$. Now let x be a complex number of absolute value 1 and set $\alpha = x\beta + \gamma$. Then since $x\bar{x} = 1$, we have

$$
\begin{aligned}
xB(\alpha, \alpha) &= xB(x\beta + \gamma, x\beta + \gamma) \\
&= x^2 \bar{x} B(\beta, \beta) + x^2 B(\beta, \gamma) + x\bar{x} B(\gamma, \beta) + x B(\gamma, \gamma) \\
&= x^2 B(\beta, \gamma) + x\big(B(\beta, \beta) + B(\gamma, \gamma)\big) + B(\gamma, \beta)
\end{aligned}
$$

Since $B(\beta, \gamma) \neq 0$, the above right hand quadratic can have at most two roots and hence $x \in \mathbb{C}$ exists with $|x| = 1$ and $B(\alpha, \alpha) \neq 0$. $\square$

We say that B is *normal* if $B(\alpha, \beta) = 0$ implies $B(\beta, \alpha) = 0$. One example is as follows. A form $B \colon V \times V \to \mathbb{C}$ is said to be *Hermitian symmetric* if for all $\alpha, \beta \in V$ we have

$$
B(\beta, \alpha) = \overline{B(\alpha, \beta)}
$$

Of course, any scalar multiple of a Hermitian symmetric form is normal. Conversely, we have

THEOREM 28.2. *Let* $B \colon V \times V \to \mathbb{C}$ *be a normal Hermitian bilinear form. Then* B *is a scalar multiple of a Hermitian symmetric form.*

A proof of this result is outlined in the first four problems of this section. Now the structure of this special type of form is easily obtained. Again we write $\alpha \perp \beta$ if $B(\alpha, \beta) = 0$.

THEOREM 28.3. *Let* V *be a finite dimensional vector space over* $\mathbb{C}$ *and let* $B \colon V \times V \to \mathbb{C}$ *be a normal Hermitian bilinear form. Then*

$$
V = W_1 + W_2 + \cdots + W_k + U
$$

is a perpendicular direct sum of the nonisotropic lines W_i *and the isotropic subspace* U.

If $\alpha = [a_{ij}] \in \mathbb{C}^{n \times n}$ we define its *conjugate transpose* α^* to be the matrix $\alpha^* = [b_{ij}]$ where $b_{ij} = \overline{a_{ij}}$. The matrix α is said to be *Hermitian symmetric* if $\alpha^* = \alpha$. We have of course

THEOREM 28.4. *Let* V *be a finite dimensional complex vector space with basis* $\mathcal{C}$ *and let* $B \colon V \times V \to \mathbb{C}$ *be a Hermitian bilinear form. Then* B *is Hermitian symmetric if and only if* $\alpha = [B]^{\mathcal{C}}$ *is a Hermitian symmetric matrix..*

Furthermore, the formula for change of basis is

THEOREM 28.5. *Let V be a finite dimensional complex vector space with bases $\mathcal{A}$ and $\mathcal{C}$ and let $B\colon V \times V \to \mathbb{C}$ be a Hermitian bilinear form. Then*

$$[B]^{\mathcal{C}} = \left(c[I]_{\mathcal{A}}\right)\cdot[B]^{\mathcal{A}}\cdot\left(c[I]_{\mathcal{A}}\right)^{*}$$

Now Example 28.1 is of course an example of a *complex inner product space*. Formally, such a space is a complex vector space V with a *complex inner product* or *complex dot product* $\alpha \bullet \beta$ defined on it and satisfying

C1. $B(\alpha, \beta) = \alpha \bullet \beta$ is a Hermitian symmetric form.
C2. $\alpha \bullet \alpha$ is real and nonnegative.
C3. $\alpha \bullet \alpha = 0$ if and only if $\alpha = 0$.

Again we define the *norm* of a vector $\alpha \in V$ to be $\|\alpha\| = \sqrt{\alpha \bullet \alpha} \geq 0$. The relationship between the inner product and the norm is not as simple here. We have

LEMMA 28.2. *Let V be a complex inner product space. If $\alpha, \beta \in V$, then*

$$\alpha \bullet \beta = \frac{1}{2}\left(\|\alpha + \beta\|^{2} - \|\alpha\|^{2} - \|\beta\|^{2}\right)$$

$$+ \frac{1}{2}\left(\|\alpha + i\beta\|^{2} - \|\alpha\|^{2} - \|\beta\|^{2}\right)i$$

where of course $i = \sqrt{-1}$.

PROOF. We have

$$\|\alpha + \beta\|^{2} = (\alpha + \beta) \bullet (\alpha + \beta) = \alpha \bullet \alpha + \alpha \bullet \beta + \beta \bullet \alpha + \beta \bullet \beta$$
$$= \|\alpha\|^{2} + 2\,\Re\mathfrak{e}(\alpha \bullet \beta) + \|\beta\|^{2}$$

Thus the real part of $\alpha \bullet \beta$ is given by

$$\Re\mathfrak{e}(\alpha \bullet \beta) = \frac{1}{2}\left(\|\alpha + \beta\|^{2} - \|\alpha\|^{2} - \|\beta\|^{2}\right)$$

In addition, we obtain the imaginary part using

$$\Im\mathfrak{m}(\alpha \bullet \beta) = \Re\mathfrak{e}\left((-i(\alpha \bullet \beta))\right) = \Re\mathfrak{e}(\alpha \bullet i\beta)$$
$$= \frac{1}{2}\left(\|\alpha + i\beta\|^{2} - \|\alpha\|^{2} - \|\beta\|^{2}\right)$$

since $\|i\beta\| = \|\beta\|$. $\square$

If V is a finite dimensional complex inner product space, then V has an orthonormal basis. In fact the Gram-Schmidt method yields a constructive proof of this result.

THEOREM 28.6. *Let V be a complex inner product space with basis $\mathcal{A} = \{\alpha_1, \alpha_2, \ldots, \alpha_n\}$. We define vectors $\gamma_1, \gamma_2, \ldots, \gamma_n$ inductively by $\gamma_1 = \alpha_1$ and*

$$\gamma_{i+1} = \alpha_{i+1} - \sum_{k=1}^{i} \frac{(\alpha_{i+1} \bullet \gamma_k)}{(\gamma_k \bullet \gamma_k)}\gamma_k$$

Then,

$$\mathcal{C} = \left\{ \frac{1}{\|\gamma_1\|}\gamma_1, \frac{1}{\|\gamma_2\|}\gamma_2, \ldots, \frac{1}{\|\gamma_n\|}\gamma_n \right\}$$

is an orthonormal basis for V.

In terms of such an orthonormal basis, the inner product is easily describable.

LEMMA 28.3. *Let V be a complex inner product space with orthonormal basis $\mathcal{C} = \{\gamma_1, \gamma_2, \ldots, \gamma_n\}$. If $\alpha = \sum_i a_i\gamma_i$ and $\beta = \sum_i b_i\gamma_i$ are vectors in V, then $a_i = \alpha \bullet \gamma_i$ and $b_i = \beta \bullet \gamma_i$ so*

$$\alpha = \sum_i (\alpha \bullet \gamma_i)\gamma_i$$

$$\alpha \bullet \beta = \sum_i a_i \bar{b}_i = \sum_i (\alpha \bullet \gamma_i)(\gamma_i \bullet \beta)$$

$$\|\alpha\|^2 = \sum_i |a_i|^2 = \sum_i |(\alpha \bullet \gamma_i)|^2$$

The Cauchy-Schwarz inequality follows from a minor modification of the original proof.

THEOREM 28.7. *Let V be a complex inner product space and let $\alpha, \beta \in V$. Then*

$$\|\alpha\|\cdot\|\beta\| \geq |\alpha \bullet \beta|$$

Moreover equality occurs if and only if α is a scalar multiple of β or β is a scalar multiple of α.

PROOF. We may assume that α and β are nonzero vectors and set $\gamma = \|\beta\|^2\alpha - (\alpha \bullet \beta)\beta$. Then

$$0 \leq \|\gamma\|^2 = (\|\beta\|^2\alpha - (\alpha \bullet \beta)\beta) \bullet (\|\beta\|^2\alpha - (\alpha \bullet \beta)\beta)$$

$$= \|\beta\|^4(\alpha \bullet \alpha) - \|\beta\|^2\overline{(\alpha \bullet \beta)}(\alpha \bullet \beta) - (\alpha \bullet \beta)\|\beta\|^2(\beta \bullet \alpha)$$

$$+ (\alpha \bullet \beta)\overline{(\alpha \bullet \beta)}(\beta \bullet \beta)$$

$$= \|\beta\|^2\cdot(\|\beta\|^2\|\alpha\|^2 - |\alpha \bullet \beta|^2)$$

and we have $\|\alpha\|^2\cdot\|\beta\|^2 \geq |\alpha \bullet \beta|^2$. Moreover if equality occurs then $\gamma = 0$ and hence α and β are multiples of each other. $\square$

An easy analog of Sylvester's Law of Inertia holds.

THEOREM 28.8. *Let V be a finite dimensional complex vector space and let $B\colon V \times V \to \mathbb{C}$ be a Hermitian symmetric bilinear form. Then there exists a basis $\mathcal{C} = \{\gamma_1, \gamma_2, \ldots, \gamma_n\}$ such that*

$$[B]^{\mathcal{C}} = \mathrm{diag}(1, 1, \ldots, 1, -1, -1, \ldots, -1, 0, 0, \ldots, 0)$$

Furthermore the number of 1's, -1's and 0's that occur on the diagonal is uniquely determined by B.

Finally let V be a complex inner product space. We say that $T\colon V \to V$ is a *Hermitian symmetric* linear transformation if for all $\alpha, \beta \in V$

$$\alpha T \bullet \beta = \alpha \bullet \beta T$$

THEOREM 28.9. *Let V be a finite dimensional inner product space over the complex numbers and let $T\colon V \to V$ be Hermitian symmetric. Then all eigenvalues of T are real and V has an orthonormal basis consisting of eigenvectors of T.*

PROOF. Let $a \in \mathbb{C}$ be an eigenvalue for T with eigenvector α. Then

$$a\alpha \bullet \alpha = \alpha T \bullet \alpha = \alpha \bullet \alpha T = \alpha \bullet a\alpha$$

so

$$a(\alpha \bullet \alpha) = \bar{a}(\alpha \bullet \alpha)$$

Since $\alpha \neq 0$ we have $a = \bar{a}$ and hence a is real. Since $\mathbb{C}$ is algebraically closed, T does have an eigenvalue. The remainder of the proof follows as in the real case. $\square$

COROLLARY 28.1. *Every Hermitian symmetric matrix is similar to a diagonal matrix with real entries.*

Problems

Let $B\colon V \times V \to \mathbb{C}$ be a normal Hermitian bilinear form. In the next four problems, we prove Theorem 28.2.

28.1. For $\sigma, \tau, \eta \subset V$, derive the identity

$$B(\eta, \sigma)\overline{B(\tau, \eta)} = \overline{B(\sigma, \eta)}B(\eta, \tau) \qquad (*)$$

(Hint. See the proof of Theorem 22.1.)

28.2. Assume that there exists $\gamma \in V$ with $B(\gamma, \gamma) > 0$ and fix this element. Setting $\tau = \eta = \gamma$, $\sigma = \beta$ in $(*)$, deduce that

$$B(\gamma, \beta) = \overline{B(\beta, \gamma)}$$

for all $\beta \in V$.

28.3. Let $\alpha, \beta \in V$ and choose $c \in \mathbb{C}$ so that $B(\gamma, c\gamma + \alpha) \neq 0$. Setting $\sigma = \beta$, $\eta = c\alpha + \gamma$, $\tau = \gamma$ in $(*)$, deduce that

$$B(\alpha, \beta) = \overline{B(\beta, \alpha)}$$

and hence that B is Hermitian symmetric.

28.4. Finish the proof of Theorem 28.2.

28.5. Verify that Example 28.1 yields a complex inner product space.

28.6. Find complex analogs for the second and third proofs of the Cauchy-Schwarz inequality.

28.7. Do the Triangle inequality and the Parallelogram law hold for complex inner product spaces?

28.8. Let $\alpha, \beta \in \mathbb{C}^{n \times n}$ and let $a \in \mathbb{C}$. Prove that

$$(\alpha + \beta)^* = \alpha^* + \beta^*$$
$$(\alpha\beta)^* = \beta^* \alpha^*$$
$$(a\alpha)^* = \bar{a}\alpha^*$$

28.9. Let V be a complex inner product space, let $T: V \to V$ be a linear transformation and let $\mathcal{C} = \{\gamma_1, \gamma_2, \ldots, \gamma_n\}$ be an orthonormal basis for V. Show that T is Hermitian symmetric if and only if the matrix $_\mathcal{C}[T]_\mathcal{C}$ is Hermitian symmetric.

28.10. Finish the proof of Theorem 28.9 and prove Corollary 28.1.

CHAPTER V

Infinite Dimensional Spaces

29. Existence of Bases

The goal now is to study infinite dimensional vector spaces, to show that bases exist and that dimension makes sense. To do this, we need a better understanding of set theory.

Of course, set theory, like all of mathematics, is based on certain axioms. Most of these are intuitive. For example, we can take unions and intersections, and we can use set builder notation as long as we are careful. The latter means that we can define a set as long as we know the objects we are putting into it. Thus we cannot look at the set of all sets because one of the objects in this set would be itself and we don't know it. Of course we can look at the collection of all sets as long as we do not view this collection as a set. Similarly, we cannot define a set A by $A = \{x \mid x \notin A\}$ since again we are trying to define A in terms of A. These sorts of constructions lead to contradictions. But as long as we are careful, set theory behaves itself.

Another thing we do is choose elements from a set. For example, if $V \neq 0$, we take an element $0 \neq \alpha \in V$. We can do this because the set $V \setminus \{0\}$ is given to be nonempty. Similarly if $B \colon V \times V \to F$ is nonzero then we can choose $\alpha, \beta \in V$ with $B(\alpha, \beta) \neq 0$. Again this follows because the set $\{(\sigma, \tau) \in V \times V \mid B(\sigma, \tau) \neq 0\}$ is nonempty. But can we make such choices simultaneously from infinitely many sets. The answer is "yes" and "no".

The "no" answer comes from the fact that basic set theory does not allow us to do it. Indeed there are so-called *models of set theory* where this kind of choice is not possible. The "yes" answer comes about because we usually add an additional axiom to basic set theory known as the *Axiom of Choice* that allows this to occur. The way the axiom is formulated is interesting. Suppose we are given a family of nonempty sets $\{A_i \mid i \in I\}$ where I is the index set or set of labels. For convenience, let $A = \bigcup_{i \in I} A_i$. Then a *choice function* is a map $f \colon I \to A$ such that $f(i) \in A_i$ for all i. Thus the existence of such a function f gives us a way to make a simultaneous choice of elements one from each A_i. The Axiom of Choice says that such choice functions exist. We state it below as a theorem.

THEOREM 29.1. *Let A be a set and let $\{A_i \mid i \in I\}$ be a family of nonempty subsets of A indexed by the set I. Then there exists a function $f \colon I \to A$ such that $f(i) \in A_i$ for all $i \in I$.*

PROOF. This is an axiom that we assume.   □

Now it is not immediately clear how useful this axiom really is. But it turns out that there are a dozen or so equivalent formulations that are more useful. Among these are: the Compactness theorem, the Hausdorff maximal principle, Kuratowski's lemma, Tukey's lemma, the Well-ordering principle, Zermelo's postulate, and Zorn's lemma. Algebraist tend to use Zorn's lemma or Well-ordering, so we will restrict our attention to these. The fact that Zorn's lemma is equivalent to the Axiom of Choice means that one can prove Zorn's lemma using the Axiom of Choice and conversely one can prove the Axiom of Choice using Zorn's lemma.

We need some definitions.

DEFINITION 29.1. We say that $(A, \leq)$ is a *partially ordered set* or a *poset* if the inequality $\leq$ is defined on A. By this we mean that for some pairs $a, b \in A$ we have $a \leq b$. Furthermore, for all $a, b, c \in A$, the inequality satisfies

PO1. $a \leq a$.
PO2. $a \leq b$ and $b \leq c$ imply $a \leq c$.
PO3. $a \leq b$ and $b \leq a$ imply $a = b$.

Of course the second condition above is the *transitive law*. As usual, we will use $b \geq a$ to mean $a \leq b$, and we will use $a < b$ or $b > a$ to mean less than or equal to, but not equal.

An element $m \in A$ is said to be *maximal* if there are no properly larger elements in A. That is, if $m \leq a$ then $m = a$. Similarly $n \in A$ is *minimal* if there are no properly smaller elements in A.

A standard example here is as follows. Let F be a set and let A be the set of all subsets of F. Then A is a partially ordered set with inequality being setwise inclusion $\subseteq$. Obviously F is the unique maximal member of A and $\emptyset$ is the unique minimal member of A.

DEFINITION 29.2. Next, $(A, \leq)$ is said to be a *linearly ordered set* or a *totally ordered set* if

LO1. $(A, \leq)$ is a partially ordered set.
LO2. For all $a, b \in A$ we have $a \leq b$ or $b \leq a$.

Thus this occurs when A is partially ordered and all pairs of elements are *comparable*. Of course both the set of rational numbers $\mathbb{Q}$ and the set of real numbers $\mathbb{R}$ are totally ordered with the usual inequality.

DEFINITION 29.3. Finally, $(A, \leq)$ is a *well-ordered set* if

WO1. $(A, \leq)$ is a linearly ordered set.
WO2. Every nonempty subset of A has a minimal member.

Notice that $\mathbb{Z}^+$ the set of positive integers is well-ordered, but $\mathbb{Q}^+$ the set of positive rationals is not. Indeed, for example, the subset $B = \{q \in \mathbb{Q}^+ \mid q > 1\}$ does not have a minimal member. Indeed, the lower bound $q = 1$ is not in B.

Suppose $(A, \leq)$ is a partially ordered set and let B be a subset of A. Then an element $u \in A$ is an *upper bound* for B if $b \leq u$ for all $b \in B$. Note that u need not be an element of B. We can now state Zorn's lemma and prove it using the Axiom of Choice. The proof of the claim below is bit tedious. The remainder of the argument is more routine.

THEOREM 29.2. *Let $(A, \leq)$ be a nonempty partially ordered set and assume that every linearly ordered subset of A has an upper bound in A. Then A has a maximal element.*

PROOF. Assume by way of contradiction that A has no maximal elements.

Let B be any linearly ordered subset of A. Then by assumption, B has an upper bound $u \in A$. Furthermore, u is not maximal, by the above, so there exists $v \in A$ with v properly larger than u, that is $u < v$. Thus for all $b \in B$ we have $b \leq u < v$ so v is a "strict" upper bound for B. We have shown that each such B has a strict upper bound. Hence by the Axiom of Choice, Theorem 29.1, there exists a choice function f from the set of linearly ordered subsets of A to A itself so that $f(B)$ is a strict upper bound for B. Note that B above is allowed to be the empty set.

Again suppose that B is a linearly ordered subset of A. If e is any element of B, then
$$B_e = \{b \in B \mid b < e\}$$
is a linearly ordered subset of B and hence of A. Clearly e is a strict upper bound for B_e, but there is no reason to believe that $e = f(B_e)$. When this happens for all such elements e then B becomes rather special. Specifically, we say that B is a Z-subset of A if

    i. B is well-ordered.
    ii. For all $e \in B$ we have $e = f(B_e)$.

Note that (i) above is a stronger assumption than B just being linearly ordered. Notice also that $\{f(\emptyset)\}$ is a Z-subset of size 1.

CLAIM. *Let $B \neq C$ be two Z-subsets. Then either $C = B_b \subseteq B$ for some $b \in B$ or $B = C_c \subseteq C$ for some $c \in C$.*

PROOF OF THE CLAIM. Since $B \neq C$ we cannot have both $B \subseteq C$ and $C \subseteq B$. By symmetry we can assume that $B \not\subseteq C$. Now B is well-ordered, so we can choose b to be a minimal element in the nonempty subset $B \setminus (B \cap C)$. Clearly minimality then implies that $B_b \subseteq C$. If $C = B_b$, then the claim is proved. If not, we can choose c to be a minimal element of the nonempty set $C \setminus B_b$. Again minimality yields

$C_c \subseteq B_b$. Note that $b \in B \setminus B_b$ so $b \in B \setminus C_c$ and hence, since B is well-ordered, we can choose d to be minimal in $B \setminus C_c$ with $d \leq b$. Again, minimality and transitivity yield $B_d \subseteq C_c$ so

$$B_d \subseteq C_c \subseteq B_b \subseteq C$$

We are given $d \leq b$. If $b = d$, then $B_d = B_b = C_c$ and hence by condition (ii), $b = f(B_b) = f(C_c) = c$, a contradiction since $b \notin C$. On the other hand, if $d < b$ then $d \in B_b \subseteq C$. But $d \notin C_c$ and C is linearly ordered so $c \leq d$. Now $d \in C$, so C_d is defined and $C_c \subseteq C_d \cap B \subseteq B_d$. Thus $B_d = C_c$ and again (ii) yields $d = f(B_d) = f(C_c) = c$ and $c \in B$. Now $c = d < b$ and $c \in B$ so $c \in B_b$ and this contradicts the definition of the element c. Thus the claim is proved. $\square$

Finally, let M be the union of all Z-subsets of A. We show that M is a Z-subset. First let $x, y \in M$ so that $x \in X$ and $y \in Y$ where X and Y are Z-subsets. By the Claim we know that X and Y are comparable sets and by symmetry we can assume that $X \subseteq Y$. Then $x, y \in Y$ and Y is linearly ordered, so we conclude that $x \leq y$ or $y \leq x$. Thus M is linearly ordered.

Next we observe that if X is a Z-subset of A and if $x \in X$, then $M_x = X_x$. Clearly $X_x \subseteq M_x$. For the other inclusion, let $y \in M_x \subseteq M$, so that $y \in Y$ for some Z-subset Y. If $Y \subseteq X$ then $y \in X$ and hence $y \in X_x$. On the other hand, if $Y \not\subseteq X$, then the Claim implies that $X = Y_t$ for some $t \in Y$. But then $x \in X = Y_t$ so $x < t$ and $y \in Y \setminus Y_t$ so $t \leq y$. Thus $x < t \leq y$ and this contradicts the fact that $y < x$.

With this, we see that M satisfies condition (ii). Indeed, let $x \in M$ and as above note that $M_x = X_x$ for some Z-set X containing x. Then $f(M_x) = f(X_x) = x$ since the Z-subset X satisfies (ii). Finally, let S be a nonempty subset of M and choose a Z-subset X with $S \cap X \neq \emptyset$. Since X is well-ordered, $S \cap X$ has a minimal element x. In particular, $S \cap X_x = \emptyset$ so $S \cap M_x = \emptyset$ and thus x is a minimal element of S. We conclude that M is a Z-subset of A and hence clearly the largest Z-subset of A.

Since M is a linearly ordered subset of A, $v = f(M)$ exists and we form $N = M \cup \{v\}$. Note that v is strictly larger than all elements of M so N is strictly larger than M. Now N is surely linearly ordered and well-ordered since M is. Furthermore, by definition of v, it is clear that N satisfies (ii). Thus N is a Z-subset of A strictly larger than M, and this contradicts the maximality of M. We conclude that our original assumption is false, so A has a maximal element and the theorem is proved. $\square$

The Well-ordering principle below says that well-orderings abound. We prove it easily using a standard Zorn's lemma argument. If A is a linearly ordered set, we say that $B \subseteq A$ is an *initial segment* if for all $a \in A$ and $b \in B$, the inequality $a \le b$ implies that $a \in B$. In other words, B is the set of "small" elements of A.

THEOREM 29.3. *Any set can be well-ordered.*

PROOF. Let A be the given set. We consider the subsets of A that can be well-ordered. Specifically, let $\mathcal{W}$, the set of all pairs $(B, \le)$ where B is a subset of A and $\le$ defines a well-ordering on B. Next we introduce an ordering on these pairs by defining $(B, \le) \preceq (C, \le)$ to mean that $B \subseteq C$, both $\le$'s agree on B, and B is an initial segment of C with respect to $\le$. Since an initial segment of an initial segment is an initial segment, it follows easily that $\mathcal{W}$ is a partially ordered set under the inequality $\preceq$.

Now suppose $\mathcal{X} = \{(X_i, \le_i) \mid i \in I\}$ is a nonempty linearly ordered subset of $\mathcal{W}$ where we have written $\le_i$ for the inequality defined on X_i. Let $X = \bigcup_{i \in I} X_i \subseteq A$. Our goal is to define an ordering $\le$ on X, to show that $(X, \le) \in \mathcal{W}$ and that $(X, \le)$ is an upper bound for $\mathcal{X}$. To start with, let $x, y, z \in X$. Then $x \in X_i$, $y \in X_j$ and $z \in X_k$ for suitable $i, j, k \in I$. Since $\mathcal{X}$ is linearly ordered, one of these three subsets is largest, say X_m, and then $x, y, z \in X_m$. In other words, any three elements of X are contained in a common X_m.

Now we define $\le$ on X. To this end, let $x, y \in X$ and let $x, y \in X_r$ for some r. Then we say that $x \le y$ if and only if $x \le_r y$. Notice that if $x, y \in X_s$ for some other s, then either $(X_r, \le_r) \preceq (X_s, \le_s)$ or $(X_s, \le_s) \preceq (X_r, \le_r)$. In either case, $\le_r$ and $\le_s$ agree on the smaller set and hence $x \le_r y$ if and only if $x \le_s y$. Thus $x \le y$ is well defined, independent of the choice of r.

Next we show that $(X_i, \le_i) \preceq (X, \le)$ for all i. To this end, note that by definition the two inequalities agree on the smaller set X_i. Next suppose $x \in X_i$ and $y \in X$ with $y \le x$. Then $y \in X_j$ for some j and either $X_j \subseteq X_i$ or $X_i \subseteq X_j$. In the latter case we know that X_i is an initial segment of X_j, so either inclusion implies that $y \in X_i$. Thus X_i is indeed an initial segment of X and $(X_i, \le_i) \preceq (X, \le)$.

Finally we show that $(X, \le) \in \mathcal{W}$ and for this we need $\le$ to be a well-ordering. Since each $(X_i, \le_i)$ is a linear ordering and since we can put three elements of X in a common X_i, it follows easily that $\le$ is a linear ordering. Thus we need only check the existence of minimal elements. To this end, let S be a nonempty subset of X and choose X_r so that $S \cap X_r \ne \emptyset$. Since X_r is well-ordered, $S \cap X_r$ has a minimal element, say x. But X_r is an initial segment of X and hence x is clearly a minimal element of S. We conclude that $(X, \leq) \in \mathcal{W}$ is an upper bound for $\mathcal{X}$ and Zorn's lemma (Theorem 29.2) implies that $\mathcal{W}$ has a maximal member $(B, \leq)$.

If $B \neq A$ we can choose $c \in A \backslash B$ and extend B to the strictly larger set $C = B \cup \{c\}$. Furthermore, we can extend the ordering on B to C by defining $b < c$ for all $b \in B$. In this way, C becomes a well-ordered set with B as an initial segment. Thus clearly $(B, \leq) \prec (C, \leq) \in \mathcal{W}$ and this contradicts the maximality of $(B, \leq)$. We conclude that $A = B$ is well-ordered. $\qquad \square$

As an easy consequence, we have

THEOREM 29.4. *The Axiom of Choice, Zorn's lemma and the Well-ordering principle are all equivalent.*

PROOF. We have already shown that the Axiom of Choice (Theorem 29.1) implies Zorn's lemma (Theorem 29.2). Furthermore, Zorn's lemma implies the Well-ordering principle (Theorem 29.3). It remains to show that Well-ordering implies the Axiom of Choice and this is basically trivial. Indeed, let A be a set and let $\{A_i \mid i \in I\}$ be a family of nonempty subsets of A indexed by the set I. Well-order the set A and observe that each nonempty subset B of A has a unique minimal element in B which we denote by $\min(B)$. Then obviously the function $f \colon I \to A$ given by $f(i) = \min(A_i)$ is a choice function. $\qquad \square$

We can now apply these set theoretic assumptions to linear algebra. Let V be a vector space over a field F. Since we can only do finite arithmetic in V, as in Section 4, we say that $L \subseteq V$ is a linearly independent subset if for all finitely many elements $\alpha_1, \alpha_2, \ldots, \alpha_n \in L$, any linear dependence is trivial. That is

$$a_1 \alpha_2 + a_2 \alpha_2 + \cdots + a_n \alpha_n = 0$$

implies that the field elements $a_1 = a_2 = \cdots = a_n = 0$. Note that the empty set $\emptyset$ is linearly independent. Similarly the subset S spans V if every element of V can be written as a finite linear combination of elements of S. That is, $V = \langle S \rangle$. Of course, a basis B is a linearly independent spanning set. As we know, if B is a basis then

    i. Every element of V can be written uniquely as a finite linear combination of elements of B.

    ii. B is a maximal linearly independent set.

    iii. B is a minimal spanning set.

The goal now is to prove that bases exist and (ii) above offers a hint as to how to prove the result. Clearly we apply Zorn's lemma.

THEOREM 29.5. *Let V be a vector space over the field F and let $L \subseteq S \subseteq V$ be given where L is a linearly independent set and where S spans V. Then there exists a basis B of V with $L \subseteq B \subseteq S$.*

PROOF. Consider the set $\mathcal{X}$ of all linearly independent subsets X of V with $L \subseteq X \subseteq S$. Then $\mathcal{X}$ is nonempty since $L \in \mathcal{X}$. Furthermore, $\mathcal{X}$ is a partially ordered set using set inclusion $\subseteq$.

Suppose $\mathcal{Y} = \{Y_i \mid i \in I\}$ is a nonempty linearly ordered subset of $\mathcal{X}$. We show that $Y = \bigcup_{i \in I} Y_i$ is an upper bound for $\mathcal{Y}$ in $\mathcal{X}$. First since $L \subseteq Y_i \subseteq S$ for all $i \in I$, we have $L \subseteq Y \subseteq S$. Next, let $\alpha_1, \alpha_2, \dots, \alpha_n$ be finitely many vectors in Y and choose subscripts $i_1, i_2, \dots, i_n \in I$ with $\alpha_j \in Y_{i_j}$. Since $\mathcal{Y}$ is linearly ordered under set inclusion, one of the subscripts, say i_m, corresponds to the largest Y_{i_j}. Then $\alpha_j \in Y_{i_j} \subseteq Y_{i_m}$ for all $j = 1, 2, \dots, n$. But Y_{i_m} is a linearly independent set and hence $\alpha_1, \alpha_2, \dots, \alpha_n$ are linearly independent vectors. We conclude that Y is a linearly independent set sandwiched between L and S, so $Y \in \mathcal{X}$. Clearly Y is an upper bound for $\mathcal{Y}$ in $\mathcal{X}$.

We have shown that every linearly ordered subset of $\mathcal{X}$ has an upper bound in $\mathcal{X}$ and hence Zorn's lemma (Theorem 29.2) implies that $\mathcal{X}$ contains a maximal member B. We show that B is a basis for V and we already know that B is linearly independent. Thus it suffices to show that B spans V.

Of course, $\langle B \rangle \supseteq B$. Next let $\gamma \in S \setminus B$. Then $B < B \cup \{\gamma\} \subseteq S$ so the maximality of B implies that $B \cup \{\gamma\}$ is linearly dependent. In other words there exists a dependence relation of the form

$$b_1 \beta_1 + b_2 \beta_2 + \cdots + b_n \beta_n + c\gamma = 0$$

with $\beta_1, \beta_2, \dots, \beta_n \in B$ and coefficients in F not all 0. If $c = 0$ then the above becomes

$$b_1 \beta_1 + b_2 \beta_2 + \cdots + b_n \beta_n = 0$$

a linear dependence relation for elements of B. But B is a linearly independent set so we have $b_1 = b_2 = \cdots = b_n = 0$, a contradiction. Thus $c \neq 0$, so we can divide by c and conclude that

$$\gamma = \frac{(-b_1)}{c}\beta_1 + \frac{(-b_2)}{c}\beta_2 + \cdots + \frac{(-b_n)}{c}\beta_n$$

and hence $\gamma \in \langle B \rangle$. Thus $\langle B \rangle \supseteq S$ and since $\langle B \rangle$ is a subspace of V we have $\langle B \rangle \supseteq \langle S \rangle$. But S spans V so $\langle S \rangle = V$ and therefore $\langle B \rangle = V$. In other words, B is a linearly independent spanning set for V and hence B is a basis for V.   □

As an immediate consequence we have

COROLLARY 29.1. *Let V be a vector space over a field F. Then*

  i. V has a basis.

 ii. Any linearly independent subset of V is contained in a basis, namely a maximal linearly independent set.

iii. Any spanning set for V contains a basis, namely a minimal spanning set.

PROOF. Take $L = \emptyset$ and $S = V$ in the previous theorem. These are linearly independent and spanning sets, respectively, so a basis exists. Next if L is given, take $S = V$ and conclude that there exists a basis $B \supseteq L$. Finally if S is given, take $L = \emptyset$ and deduce that a basis B exists with $B \subseteq S$. □

It remains to show that all bases for V have the same "size", whatever that means, and we consider this in the next section.

Problems

Let $(A, \leq)$ be a partially ordered set.

29.1. Show that the following two conditions are equivalent.

  i. A satisfies the *maximal condition (MAX)*, which means that every nonempty subset of A has a maximal member.

 ii. A satisfies the *ascending chain condition (ACC)*, namely A has no strictly ascending chain

$$a_1 < a_2 < \cdots < a_n < \cdots$$

indexed by the positive integers.

29.2. Show that the following two conditions are equivalent.

  i. A satisfies the *minimal condition (MIN)*, which means that every nonempty subset of A has a minimal member.

 ii. A satisfies the *descending chain condition (DCC)*, namely A has no strictly descending chain

$$a_1 > a_2 > \cdots > a_n > \cdots$$

indexed by the positive integers.

29.3. Show that A is well-ordered if and only if every nonempty subset has a unique minimal member. (Hint. To prove that the ordering is linear consider subsets of size 2.)

29.4. Define the relation $\preceq$ on A by $a \preceq b$ if and only if $b \leq a$. Show that $(A, \preceq)$ is also a partially ordered set that has the effect of interchanging maximal and minimal and also upper bound and lower bound. Use this to translate Zorn's lemma into a result guaranteeing the existence of a minimal member of A.

29.5. Take a close look at the proof of Theorem 29.2. Did we actually prove the slightly stronger result, namely if all well-ordered subsets of A have upper bounds in A, then A has a maximal element.

29.6. Let $B \subseteq A$. We say that B is a *chain* if it is linearly ordered, that is if all elements of B are comparable. In the other extreme, B is said to be an *anti-chain* if no distinct elements are comparable. Prove that A has a maximal chain and a maximal anti-chain.

Now let V be a vector space over F.

29.7. Let $(A, \leq)$ be the partially ordered set of subspaces of V with inequality corresponding to set inclusion $\subseteq$. If V has an infinite linearly independent subset, show that A does not satisfy ACC or DCC. (Hint. As we will see in Proposition 30.1, V has a linearly independent subset $\{\alpha_1, \alpha_2, \ldots, \alpha_n, \ldots\}$ indexed by the positive integers $\mathbb{Z}^+$.)

29.8. Verify that bases satisfy the three conditions listed just before Theorem 29.5.

29.9. Find an example of V and a descending chain
$$S_1 \supseteq S_2 \supseteq \cdots \supseteq S_n \supseteq \cdots$$
of spanning sets of V such that the intersection $S = \bigcap_{n=1}^{\infty} S_n$ does not span V. (Hint. Surprisingly we can take V to be 1-dimenional.)

29.10. Let W be a subspace of V. Show that there exists a subspace U of V with $V = W \oplus U$. (Hint. Start with a basis for W.)

30. Existence of Dimension

One can define the dimension of an infinite dimensional vector space but as we will see, it is really less useful than in the finite dimensional case. Obviously, we start by first understanding what the size of a basis might mean.

Let A be a finite set and observe that we find $|A| = n$ by counting the elements of A. We say "first element, second element, ..., nth element". Now what we are actually doing is setting up a one-to-one correspondence between the elements of A and the elements of the set $\{1, 2, \ldots, n\}$. In fact, that correspondence naturally appears when we write $A = \{a_1, a_2, \ldots, a_n\}$. Indeed, $i \leftrightarrow a_i$ is precisely the correspondence. Furthermore, if $|B| = n$ and $B = \{b_1, b_2, \ldots, b_n\}$ then we have a one-to-one correspondence between the elements of A and of B given by $a_i \leftrightarrow b_i$. Let us formalize and extend these definitions.

Let A and B be sets. Then the map $f\colon A \to B$ is a *one-to-one correspondence* if f is one-to-one and onto. Clearly this occurs if and only if there exists a back map $f^{-1}\colon B \to A$ such that $f^{-1}f = 1_A$ and $ff^{-1} = 1_B$ where 1_A is the identity map on A and 1_B is the identity map on B. We write $A \sim B$ if the maps f and f^{-1} exist.

LEMMA 30.1. *With the above notation we have*

  i. $A \sim A$ *for all* A *(reflexive law)*
  ii. $A \sim B$ *implies* $B \sim A$ *(symmetric law)*
  iii. $A \sim B$ *and* $B \sim C$ *imply* $A \sim C$ *(transitive law)*
  iv. *Suppose* $A = \bigcup_{i \in I} A_i$ *and* $B = \bigcup_{i \in I} B_i$ *are disjoint unions with* $A_i \sim B_i$ *for all* $i \in I$. *Then* $A \sim B$.

PROOF. For (i) we use $1_A\colon A \to A$ and (ii) follows since a one-to-one correspondence yields maps in both directions. Since the composition of two one-to-one correspondences is a one-to-one correspondence, (iii) is immediate. Finally, for (iv) we define a map from A to B by defining it on each of the disjoint subsets A_i. In particular, if I is finite or if there is a canonical choice for the maps $f_i\colon A_i \to B_i$, then the result is clear. In the general case, we need the Axiom of Choice to choose a fixed one-to-one correspondence f_i for each $i \in I$.   □

As in Chapter 10, the first three rules above tell us that $\sim$ is an equivalence relation. We then say that sets A and B have the same size if and only if $A \sim B$. In some sense, the *size* of A, namely $|A|$, is the collection of all sets B with $A \sim B$ namely the equivalence class of A. Thus sizes don't really have names except in a few special cases. For example, as above, A has size n if $A \sim \{1, 2, \ldots, n\}$. Next A is *countably infinite* if $A \sim \mathbb{N}$ where $\mathbb{N} = \mathbb{Z}^+ = \{1, 2, \ldots, n, \ldots\}$ is the set of natural numbers and A has the size of the *continuum* if $A \sim \mathbb{R}$.

Again, let A and B be sets. We need two more set-theoretic definitions. First, we want size to be compatible with set inclusion. Thus we say $|B| \leq |A|$ if and only if there exists a one-to-one map $f \colon B \to A$. When this occurs then f gives rise to a one-to-one correspondence between B and $f(B)$ and hence $B \sim f(B) \subseteq A$. It is clear that the inequality $|B| \leq |A|$ depends only upon the size of the sets and not on the sets themselves.

Next, we define $|A| + |B|$ to be the size of $A \cup B$ if A and B are disjoint. More generally, let $A' = \{a' \mid a \in A\}$ and $B^* = \{b^* \mid b \in B\}$ so that $A \sim A'$ and $B \sim B^*$. Furthermore, A' and B^* are disjoint because the elements of A' end in ′ while the elements of B^* end in *, and thus we can define $|A| + |B| = |A' \cup B^*|$. It follows from parts (*iii*) and (*iv*) of the previous lemma that if $A \sim \overline{A}$ and $B \sim \overline{B}$, then $A' \cup B^* \sim \overline{A}' \cup \overline{B}^*$ and hence that $|A| + |B|$ only depends upon $|A|$ and $|B|$. Furthermore, it is clear that $|A| + |B| = |B| + |A|$ and $|A| \leq |A| + |B|$. Note that the latter makes sense since $|A| + |B|$ is the size of a set.

It is interesting to observe some elementary properties of the set $\mathbb{N}$. Surprisingly part (*iii*) below uses the Axiom of Choice.

LEMMA 30.2. *Let $\mathbb{N}$ denote the set of natural numbers.*

    i. If F is a finite set then $|F| + |\mathbb{N}| = |\mathbb{N}|$.
   ii. $|\mathbb{N}| + |\mathbb{N}| = |\mathbb{N}|$.
  iii. Let $A = \bigcup_{n \in \mathbb{N}} A_n$ be a countable union of the finite sets A_n. Then either A is finite of $A \sim \mathbb{N}$.

PROOF. (*i*) If $|F| = n$, then $F \sim \{1, 2, \ldots, n\}$, and clearly $\mathbb{N} \sim \{n+1, n+2, \ldots\}$. Thus $|F| + |\mathbb{N}|$ is the size of the set

$$\{1, 2, \ldots, n\} \cup \{n+1, n+2, \ldots\} = \mathbb{N}$$

(*ii*) Now $\mathbb{N} \sim \mathbb{E}$, where $\mathbb{E}$ is the set of even positive integers and $\mathbb{N} \sim \mathbb{O}$ where $\mathbb{O}$ is the set of odd positive integers. Thus $|\mathbb{N}| + |\mathbb{N}|$ is the size of the set $\mathbb{E} \cup \mathbb{O} = \mathbb{N}$.

(*iii*) For each n, let $B_n = A_1 \cup A_2 \cup \cdots \cup A_n$ so that $A = \bigcup_n B_n$. Furthermore, the B_n's form an ascending chain of finite sets starting with $B_0 = \emptyset$. If $s_n = |B_n|$, then $B_n \setminus B_{n-1} \sim (s_{n-1}, s_n]$, where the latter is the half-open interval $\{x \in \mathbb{N} \mid s_{n-1} < x \leq s_n\}$. If the s_n's are bounded, then A is finite. Otherwise

$$A = \dot{\bigcup}_{n \in \mathbb{N}} B_n \setminus B_{n-1} \sim \dot{\bigcup}_{n \in \mathbb{N}} (s_{n-1}, s_n] = \mathbb{N}$$

where $\dot{\bigcup}$ as usual indicates disjoint union.   □

We again use the Axiom of Choice in the next construction to obtain an extension of the above result.

PROPOSITION 30.1. *Suppose A is any infinite set. Then A contains a copy of $\mathbb{N}$. Consequently, $|\mathbb{N}| \leq |A|$ and we have*

    i. If F is a finite set then $|F| + |A| = |A|$.
    ii. $|\mathbb{N}| + |A| = |A|$.

PROOF. If B is any finite subset of A, then $A \setminus B$ is nonempty. Hence, by the Axiom of Choice, there exists a choice function f from the set of all finite subsets of A to A itself such that $f(B) \in A \setminus B$. With this and mathematical induction, we define a sequence of distinct elements $a_n \in A$ by $a_1 = f(\emptyset)$ and $a_{n+1} = f(\{a_1, a_2, \ldots, a_n\})$. Clearly $\mathbb{N} \sim \{a_1, a_2, \ldots, a_n, \ldots\} \subseteq A$ and hence $|\mathbb{N}| \leq |A|$.

Now let N denote this copy of $\mathbb{N}$ in A. Then A is the disjoint union $A = N \,\dot\cup\, (A \setminus N)$. In particular, if F is any finite set disjoint from A, then the previous lemma yields

$$F \,\dot\cup\, A = (F \,\dot\cup\, N) \,\dot\cup\, (A \setminus N) \sim N \,\dot\cup\, (A \setminus N) = A$$

and similarly if $M \sim \mathbb{N}$ is disjoint from A then $M \,\dot\cup\, N \sim N$ implies

$$M \,\dot\cup\, A = (M \,\dot\cup\, N) \,\dot\cup\, (A \setminus N) \sim N \,\dot\cup\, (A \setminus N) = A$$

as required. $\qquad\qquad\qquad\qquad\qquad\qquad\qquad\qquad\qquad\qquad$ $\square$

The following result is the Schroeder Bernstein theorem.

THEOREM 30.1. *Let A and B be sets with $|A| \leq |B|$ and $|B| \leq |A|$. Then $|A| = |B|$.*

PROOF. We can better appreciate this proof if we think pictorially. To start with, $|A| \leq |B|$ means that there exists a one-to-one function $f \colon A \to B$. Similarly $|B| \leq |A|$ yields a one-to-one function $g \colon B \to A$. Since these functions are one-to-one, there exist back maps which we write as $f^{-1} \colon B \to A$ and $g^{-1} \colon A \to B$ with the understanding that f^{-1} is only defined on $f(A) \subseteq B$ and g^{-1} is only defined on $g(B) \subseteq A$. Note that $f^{-1}f = 1_A$ and $ff^{-1} = 1_{f(A)}$. Similarly we have $g^{-1}g = 1_B$ and $gg^{-1} = 1_{g(B)}$.

We can of course assume that A and B are disjoint. Our goal is to partition $A \cup B$ into a family of "horizontal lines" H with the property that $H \cap A \sim H \cap B$. To start with, take $a_0 \in A$ and move to the right by applying f, then g, then f, $\ldots$, as in the diagram below. Or we could have started with $b_0 \in B$ and moved to the right by applying g, then f, then g, $\ldots$, again as in the diagram below.

$$\xrightarrow{\ g\ } A \xrightarrow{\ f\ } B \xrightarrow{\ g\ } A \xrightarrow{\ f\ } B \xrightarrow{\ g\ } A \xrightarrow{\ f\ }$$

We can certainly continue infinitely in this direction.

Next we move to the left from a_0 by applying g^{-1}, then f^{-1}, then $g^{-1}, \ldots$, as in the diagram below. Or we could move to the left from b_0 by first applying f^{-1}, then g^{-1} and then $f^{-1} \ldots$, again as in the diagram. In either case, we can keep going until we are "blocked", that is until we reach an element of A not in the domain of g^{-1} or an element of B not in the domain of f^{-1}.

$$\xleftarrow{\;g^{-1}\;} A \xleftarrow{\;f^{-1}\;} B \xleftarrow{\;g^{-1}\;} A \xleftarrow{\;f^{-1}\;} B \xleftarrow{\;g^{-1}\;} A \xleftarrow{\;f^{-1}\;}$$

Notice that the elements to the right are all in $f(A)$ or $g(B)$ and thus can be moved to the left, eventually winding up where we started. Similarly, we can certainly apply the functions f and g alternately to the elements on the left and again return to the starting point. With this, it is clear that any element in the horizontal line H completely determines H and hence that different horizontal lines are disjoint. In other words, $A \cup B$ is partitioned by the family of horizontal lines.

Now any of these horizontal lines certainly extends infinitely to the right. But there are three different behaviors on the left. First, there are the lines that extend infinitely to the left and we call these the *unblocked* lines. The remaining ones are blocked and are either *A-blocked* if their left most member is in A or *B-blocked* if their left most member is in B. Note that the elements in the blocked lines are all distinct since they are each at a different distance from the blocking member. On the other hand, it is possible for the unblocked lines to contain only finitely many distinct elements.

It remains to show that $H \cap A \sim H \cap B$ for any such H and this is clear pictorially. If H is A-blocked, the required one-to-one correspondence $H \cap A \to H \cap B$ is a single shift to the right, namely the function f. If H is B-blocked, we use a single shift to the left, namely g^{-1}. Finally, if H is unblocked there are choices, but we again take f, the single shift to the right. By combining these, we construct a one-to-one correspondence $h \colon A \to B$ by $h(a) = g^{-1}(a)$ if a determines a B-blocked line and $h(a) = f(a)$ otherwise. This completes the proof. $\square$

Next we need two applications of Zorn's lemma with proofs that are similar but somewhat easier than the proof of the Well-ordering principle (Theorem 29.3). The first part shows that all set sizes are comparable. Of course, once we know this, then it makes sense to speak about the maximum of two such sizes.

THEOREM 30.2. *Let A and B be any two sets.*

    i. Either $|A| \leq |B|$ or $|B| \leq |A|$.

ii. If at least one of A or B is infinite then

$$|A| + |B| = \max\{|A|, |B|\}$$

PROOF. (i) Consider the set $\mathfrak{C}$ of all ordered pairs (C, f) where $C \subseteq A$ and $f \colon C \to B$ is a one-to-one function. We write $(C, f) \preceq (D, g)$ if and only if $C \subseteq D$ and both functions agree on the smaller set C. In this way, $\mathfrak{C}$ becomes a partially ordered set with unique smallest element $C = \emptyset$ and the obvious embedding $C \to B$. Now suppose $\mathfrak{L} = \{(C_i, f_i) \mid i \in L\}$ is a linearly ordered subset of $\mathfrak{C}$. Then we obtain an upper bound for $\mathfrak{L}$ by taking $C = \bigcup_{i \in L} C_i \subseteq A$ and $f \colon C \to B$ to agree with f_i on C_i. It is easy to see that f is a well-defined, one-to-one function. It now follows from Zorn's lemma that $\mathfrak{C}$ has a maximal member (M, g).

If $M = A$, then $g \colon A \to B$ is one-to-one and $|A| \le |B|$. On the other hand, if $g(M) = B$, then $g^{-1} \colon B \to A$ is one-to-one and $|B| \le |A|$. Finally if $M < A$ and $g(M) < B$, then we can choose $a \in A \setminus M$, $b \in B \setminus g(M)$ and extend the function g to $\overline{M} = M \cup \{a\}$ by defining $\bar{g}(a) = b$. But then $(\overline{M}, \bar{g}) \in \mathfrak{C}$ is strictly larger than (M, g), a contradiction.

(ii) Suppose A is an infinite set. We first show that $|A| + |A| = |A|$. To this end, recall that we have maps $' \colon A \to A'$ and $^* \colon A \to A^*$ and that $|A| + |A| = |A' \cup A^*|$. Now consider the set $\mathfrak{C}$ of all pairs (C, f) where $C \subseteq A$ and $f \colon C' \cup C^* \to C$ is a one-to-one correspondence. Again we define $(C, f) \preceq (D, g)$ if and only if $C \subseteq D$ and f and g agree on the smaller set $C' \cup C^*$. In this way, $\mathfrak{C}$ becomes a partially ordered set with unique smallest element corresponding to $C = \emptyset$. If $\mathfrak{L} = \{(C_i, f_i) \mid i \in L\}$ is a linearly ordered subset of $\mathfrak{C}$, then we obtain an upper bound for $\mathfrak{L}$ by taking $C = \bigcup_{i \in L} C_i \subseteq A$ and $f \colon C' \cup C^* \to C$ to agree with f_i on $C_i' \cup C_i^*$. It is easy to see that f is a well-defined, one-to-one correspondence. It now follows from Zorn's lemma that $\mathfrak{C}$ has a maximal member (M, g).

Note that $(M, g) \in \mathfrak{C}$ implies that $|M| + |M| = |M|$. If $A \setminus M$ is infinite, then by Proposition 30.1, $A \setminus M$ has a copy N of the natural numbers $\mathbb{N}$. But then we can extend g to the set $\overline{M}' \cup \overline{M}^*$ corresponding to $\overline{M} = M \cup N$ using the one-to-one correspondence between $N' \cup N^*$ and N given by Lemma 30.2. Of course, this contradicts the maximality of (M, g). Thus $A \setminus M$ is finite, M is infinite and $|M| = |A|$ by Proposition 30.1. It then follows that

$$|A| + |A| = |M| + |M| = |M| = |A|$$

as required.

Finally, suppose $|B| \leq |A|$. Then

$$|A| \leq |A| + |B| \leq |A| + |A| = |A|$$

so the Schroeder-Bernstein theorem (Theorem 30.1) implies that

$$|A| + |B| = |A| = \max\{|A|, |B|\}$$

and the theorem is proved.   □

The goal now is to use this machinery to show that all bases of a vector space have the same size. To start with, we need

LEMMA 30.3. *Let V be a nonzero vector space and let B and $\bar{B}$ be bases for V. Then there exist nonempty subsets C of B and $\bar{C}$ of $\bar{B}$ such that $C \sim \bar{C}$ and $\langle C \rangle = \langle \bar{C} \rangle$.*

PROOF. Since $\bar{B}$ is a basis of V, each $\beta \in B$ can be written uniquely as a finite linear combination, with nonzero coefficients, of elements of $\bar{B}$. Let us denote this finite subset of $\bar{B}$ by $\bar{B}(\beta)$, the *support* of β in $\bar{B}$. Similarly each $\bar{\beta} \in \bar{B}$ has a unique finite support $B(\bar{\beta}) \subseteq B$. Clearly $\beta \in \langle \bar{B}(\beta) \rangle$ and $\bar{\beta} \in \langle B(\bar{\beta}) \rangle$.

We now construct a sequence of finite subsets C_n of B and $\bar{C}_n$ of $\bar{B}$ as follows. To start with, choose $\beta_1 \in B$ and take $C_1 = \{\beta_1\}$. Similarly, $\bar{C}_1 = \{\bar{\beta}_1\}$ for some $\bar{\beta}_1 \in \bar{B}$. Now suppose that we have finite sets $C_n \subseteq B$ and $\bar{C}_n \subseteq \bar{B}$. Then we let $C_{n+1} = \bigcup B(\bar{\beta})$ where the union is over the finitely many elements $\bar{\beta} \in \bar{C}_n$ and similarly $\bar{C}_{n+1} = \bigcup \bar{B}(\beta)$ where the union is over the finitely many elements $\beta \in C_n$. Clearly $C_n \subseteq \langle \bar{C}_{n+1} \rangle$ and $\bar{C}_n \subseteq \langle C_{n+1} \rangle$.

Finally let $C = \bigcup_{n=1}^{\infty} C_n$ and $\bar{C} = \bigcup_{n=1}^{\infty} \bar{C}_n$. Then

$$W = \langle C \rangle = \langle \bar{C} \rangle$$

so C and $\bar{C}$ are bases of W since they span and since they are both linearly independent subsets of the corresponding bases B and $\bar{B}$. If either C or $\bar{C}$ is finite, then W is finite dimensional and Theorem 5.2 implies that $C \sim \bar{C}$. On the other hand if both C and $\bar{C}$ are infinite, then part (*iii*) of Lemma 30.2 yields $C \sim \mathbb{N} \sim \bar{C}$.   □

As a consequence of the above and Zorn's lemma, we obtain

THEOREM 30.3. *Let V be a vector space over the field F. Then any two bases of V have the same size.*

PROOF. Let B and $\bar{B}$ be bases of V. We consider the family $\mathfrak{C}$ of all pairs (C, f) where $C \subseteq B$, $f \colon C \to \bar{B}$ is a one-to-one map and where $\langle C \rangle = \langle f(C) \rangle$. As usual we write $(C, f) \preceq (D, g)$ if $C \subseteq D$ and if f

and g agree on the smaller set C. Then $\mathfrak{C}$ becomes a partialy ordered set under $\preceq$ and $\mathfrak{C}$ has the unique minimum member given by $C = \emptyset$.

Suppose $\mathfrak{L} = \{(C_i, f_i) \mid i \in L\}$ is a linearly ordered subset of $\mathfrak{C}$. Then it is easy to see that $\mathfrak{L}$ has an upper bound (C, f) where $C = \bigcup C_i$ and where $f\colon C \to \overline{B}$ agrees with f_i on C_i. We can now apply Zorn's lemma (Theorem 29.2) to conclude that $\mathfrak{C}$ has a maximal member (M, g) and we set $\overline{M} = g(M) \subseteq \overline{B}$. Since $(M, g) \in \mathfrak{C}$ it follows that $W = \langle M \rangle = \langle \overline{M} \rangle$ and $M \sim \overline{M}$. If $W = V$, then since B and $\overline{B}$ are minimal spanning sets of V, we conclude that $M = B$, $\overline{M} = \overline{B}$ and therefore $B \sim \overline{B}$.

On the other hand, suppose by way of contradiction that $W \neq V$ and, using Theorem 10.2, let $T\colon V \to V/W$ be the natural linear transformation onto the quotient space V/W with kernel W. Since $W = \langle M \rangle = \langle \overline{M} \rangle$, it follows easily that T is one-to-one on $B \setminus M$ and on $\overline{B} \setminus \overline{M}$. Furthermore, the images $T(B \setminus M)$ and $T(\overline{B} \setminus \overline{M})$ are both bases for the space V/W. (Note that, to be consistent with our recent function notation, we are writing T on the left here.)

Applying the preceding lemma to these two bases for V/W, we conclude that there exist nonempty subsets C of $B \setminus M$ and $\overline{C}$ of $\overline{B} \setminus \overline{M}$ such that $T(C) \sim T(\overline{C})$ and $\langle T(C) \rangle = \langle T(\overline{C}) \rangle$. Properties of T then imply that $C \sim \overline{C}$. Furthermore, by considering complete inverse images of subspaces, we see that $\langle M \cup C \rangle = \langle \overline{M} \cup \overline{C} \rangle$. Finally using $C \sim \overline{C}$, we see that the one-to-one correspondence $g\colon M \to \overline{M}$ extends to a one-to-one correspondence $h\colon M \cup C \to \overline{M} \cup \overline{C}$. But then $(M \cup C, h) \in \mathfrak{C}$ is properly larger than (M, g), contradicting the maximality of (M, g). This completes the proof. $\square$

We can now, of course, define the *dimension* of V to be the common size of all bases of V. We mention just one consequence of this definition and of some of the theorems above.

COROLLARY 30.1. *Let V be an infinite dimensional vector space over the field F and let $T\colon V \to W$ be a linear transformation. Then*

$$\dim_F V = \dim_F \ker T + \dim_F \operatorname{im} T$$
$$= \max\{\dim_F \ker T,\ \dim_F \operatorname{im} T\}$$

PROOF. Let A be a basis for $\ker T$. Then A is a linearly independent subset of V and hence, by Corollary 29.1, A extends to the disjoint union $A \mathbin{\dot{\cup}} B$, a basis for V. Then clearly $T\colon B \to T(B)$ is a one-to-one correspondence and $T(B)$ is a basis for $\operatorname{im} T = T(V)$. Thus

$$\dim V = |A \mathbin{\dot\cup} B| = |A| + |B| = |A| + |T(B)|$$
$$= \dim \ker T + \dim \operatorname{im} T$$

The result follows from Theorem 30.2.   □

Since infinite dimensional vector spaces can have proper subspaces of the same dimension, it is clear that dimension is a less useful tool in this infinite context.

Problems

30.1. Let $V \neq 0$ be a finite dimensional vector space over $\mathbb{Q}$. Show that $V \sim \mathbb{N}$. (Hint. Let $\{\gamma_1, \gamma_2, \ldots, \gamma_k\}$ be a basis for V and for each natural number n let V_n be the set of all vectors

$$\frac{a_1}{b_1}\gamma_1 + \frac{a_2}{b_2}\gamma_2 + \cdots + \frac{a_k}{b_k}\gamma_k$$

with $a_i, b_i \in \mathbb{Z}$ and $|a_i|, |b_i| \leq n$ for all i. Now apply Lemma 30.2.)

30.2. If A is a set, then its *power set* $\mathcal{P}(A)$ is the set of all subsets of A. Observe that $A \sim \bar{A}$ implies that $\mathcal{P}(A) \sim \mathcal{P}(\bar{A})$. Now prove that that $|A| \neq |\mathcal{P}(A)|$. (Hint. If $f\colon A \to \mathcal{P}(A)$ is given, define a subset B of A so that $b \in B$ if and only if $b \notin f(b)$. Is B in the image of f?)

30.3. Show that $|\mathbb{R}| \leq |\mathcal{P}(\mathbb{Q})| = |\mathcal{P}(\mathbb{N})|$. (Hint. Each real number r determines a "cut" $\{q \in \mathbb{Q} \mid r < q\}$ in the rational line.)

30.4. Conversely show that $|\mathcal{P}(\mathbb{N})| \leq |\mathbb{R}|$ and deduce that equality occurs. (Hint. We construct a map f from $\mathcal{P}(\mathbb{N})$ to $\mathbb{R}$, or actually to elements in the half open interval $[0, 1)$, as follows. For each subset B of $\mathbb{N}$, let $f(B)$ be the real number whose nth decimal digit, after the decimal point, is equal to 1 if $n \in B$ and equal to 0 otherwise. Thus

$$f(B) = \sum_{n \in B} \frac{1}{10^n}$$

What happens if the 10 in the denominator is replaced by 3 or by 2?)

30.5. Both $\mathbb{R}$ and $\mathbb{C}$ are vector spaces over $\mathbb{Q}$. Prove that they both have the same $\mathbb{Q}$-dimension. (Hint. First argue that these dimensions are infinite.)

30.6. Let $f\colon \mathbb{R} \to \mathbb{R}$ satisfy $f(x+y) = f(x) + f(y)$ for all $x, y \in \mathbb{R}$. If f is a continuous function prove that $f(x) = ax$ for $a = f(1) \in \mathbb{R}$. Show that there are numerous discontinuous examples.

30.7. Explain how Proposition 30.1 is actually used in the solution to Problems 26.7–26.10. Compare those arguments to the simpler solution of Problem 30.2 that is suggested above.

30.8. Consider the proof of Theorem 30.1, the Schroeder-Bernstein theorem, and give an example of a situation where the unblocked horizontal lines have only finitely many distinct elements.

30.9. The proofs of both parts of Theorem 30.2 are a bit skimpy. Carefully verify the fact that the linearly ordered sets $\mathfrak{L}$ have upper bounds in $\mathfrak{C}$.

30.10. Verify the material in the proof of Theorem 30.3 concerning the linear transformation $T\colon V \to V/W$.

Index

Printed in the United States
by Baker & Taylor Publisher Services